Springer Series on
SIGNALS AND COMMUNICATION TECHNOLOGY

SIGNALS AND COMMUNICATION TECHNOLOGY

Circuits and Systems Based
on Delta Modulation
D. G. Zrilic
ISBN 3-540-23751-8

Functional Structures in Networks
AMLn – A Language for Model Driven
Development of Telecom Systems
T. Muth
ISBN 3-540-22545-5

Radio Wave Propagation
for Telecommunication Applications
H. Sizun
ISBN 3-540-40758-8

Electronic Noise and Interfering Signals
Principles and Applications
G. Vasilescu
ISBN 3-540-40741-3

DVB
The Family of International Standards
for Digital Video Broadcasting, 2nd ed.
U. Reimers
ISBN 3-540-43545-X

Digital Interactive TV and Metadata
Future Broadcast Multimedia
A. Lugmayr, S. Niiranen, and S. Kalli
ISBN 0-387-20843-7

Adaptive Antenna Arrays
Trends and Applications
S. Chandran (Ed.)
ISBN 3-540-20199-8

Digital Signal Processing
with Field Programmable Gate Arrays
U. Meyer-Baese
ISBN 3-540-21119-5

Neuro-Fuzzy and Fuzzy-Neural Applications
in Telecommunications
P. Stavroulakis (Ed.)
ISBN 3-540-40759-6

SDMA for Multipath Wireless Channels
Limiting Characteristics and Stochastic Models
I.P. Kovalyov
ISBN 3-540-40225-X

Digital Television
A Practical Guide for Engineers
W. Fischer
ISBN 3-540-01155-2

Multimedia Communication Technology
Representation, Transmission
and Identification of Multimedia Signals
J.R. Ohm
ISBN 3-540-01249-4

Information Measures
Information and its Description in Science
and Engineering
C. Arndt
ISBN 3-540-40855-X

Processing of SAR Data
Fundamentals, Signal Processing,
Interferometry
A. Hein
ISBN 3-540-05043-4

Chaos-Based Digital Communication Systems
Operating Principles, Analysis Methods,
and Performance Evaluation
F.C.M. Lau and C.K. Tse
ISBN 3-540-00602-8

Adaptive Signal Processing
Applications to Real-World Problems
J. Benesty and Y. Huang (Eds.)
ISBN 3-540-00051-8

Multimedia Information Retrieval
and Management
Technological Fundamentals
and Applications
D. Feng, W.C. Siu, and H.J. Zhang (Eds.)
ISBN 3-540-00244-8

Structured Cable Systems
A.B. Semenov, S.K. Strizhakov,
and I.R. Suncheley
ISBN 3-540-43000-8

UMTS
The Physical Layer of the Universal Mobile
Telecommunications System
A. Springer and R. Weigel
ISBN 3-540-42162-9

Advanced Theory of Signal Detection
Weak Signal Detection
in Generalized Observations
I. Song, J. Bae, and S.Y. Kim
ISBN 3-540-43064-4

Wireless Internet Access over GSM and UMTS
M. Taferner and E. Bonek
ISBN 3-540-42551-9

Ute Jekosch

Voice and Speech Quality Perception

Assessment and Evaluation

With 34 Figures

 Springer

Dr. Ute Jekosch
Rensselaer Polytechnic Institute
School of Architecture
12180 Troy, N.Y.
USA

ISBN 10 3-540-24095-0 **Springer Berlin Heidelberg New York**
ISBN 13 978-3-540-24095-2 **Springer Berlin Heidelberg New York**

Library of Congress Control Number: 2005926886

Springer is a part of Springer Science+Business Media

springeronline.com

© Springer-Verlag Berlin Heidelberg 2005
Printed in Germany

Typesetting: Data conversion by author.
Final processing by PTP-Berlin Protago-TEX-Production GmbH, Germany
Cover-Design: design & production GmbH, Heidelberg
Printed on acid-free paper   62/3141/Yu - 5 4 3 2 1 0

Preface

As part of the steady progress being made in the field of information and telecommunication techniques, voice and speech quality assessment of systems has gained in importance over the last years. An engineering approach to voice and speech quality of systems includes the consideration of how a system is perceived by its users, and how the needs and expectations of the users develop. Thus, quality assessment is closely linked with quality prediction, and both have to take the relevant human perception and judgment factors into account. Speech quality assessments are expected to fulfil certain requirements – they must make reliable and meaningful data available as quickly and cheaply as possible. Although significant progress has been made in several areas concerned with voice and speech quality within the last two decades, there is still neither consensus on the definition of voice and speech quality and their contributing components nor on appropriate assessment and evaluation methods.

Usually, in everyday situations quality of systems becomes an issue only when expectations are not met during system use. The explicit process of judging voice or speech quality is provoked when form, content or communication characteristics of speech come to the fore: the speech signal as a carrier of information is catching attention so that quality is reflected upon.

Engineers of speech devices such as information and telecommunication systems have a vivid interest to know how users will judge system quality even before new technology is offered to the market. Thus, one important aim is to obtain findings on how speech quality assessment can be performed in a controlled artificial way, e.g. in a laboratory, without these judgments losing meaningfulness when compared to accidental judgments in natural environments.

The supporting scientific field is »voice and speech quality perception«. It is oriented towards the processes determining speech quality both in natural and artificial situations. The most pressing aim is to examine and understand the processes of speech perception and assessment, gathering and including basic knowledge pertinent to psycho-acoustics, to the theory of perception, to metrology and linguistics.

In this book the emphasis is placed on voice and speech quality assessment of systems in artificial situations. Here the main purpose is to add to knowledge so as to be able to construct artificial assessment scenarios that are as close to reality as possible with regard to the resulting voice or speech quality judgment. Thus, the basic requirement is that these scenarios should be constructed in such a way that the listeners' reactions in communicatively »sterile« circumstances correspond to their reaction in a comparable natural (therefore extrinsically uncontrolled) situation. Understanding the processes that happen during uncontrolled voice and speech quality assessments is the prerequisite for creating artificial assessment scenarios.

Therefore the objective of voice and speech quality perception as a scientific field is to provide the knowledge to describe, structure, construct and execute processes that are essential to voice and speech quality assessment. Very varied approaches to do this are used: Apart from observing, describing and analyzing natural events, one effective method is to deliberately manipulate identified or suspected factors of influence. In this way assessment scenarios are constructed in clearly defined and reproducible circumstances in a laboratory to increase knowledge of components and links. A number of various experiments is introduced and discussed. These experiments were conducted for the German language. Which is no drawback, because what is of major interest here is not the results themselves but their indicating and revealing basic processes of voice and quality quality perception. It can be assumed here that such processes are language independent to a certain extent.

This book is directed towards communication and information technology. It lies in the interdisciplinary space between communication acoustics, communication and information theory as well as linguistics. One of the aims is to work out the status of auditory assessment and instrumental measuring methods in the context of advanced speech-related psycho-acoustics. Positioning speech technologies in their various contexts highlights the need to discuss speech quality assessment and to collect evidence that will justify the approach introduced here.

I would like to thank all who have contributed to this piece of work, specifically Susanne Krause and Bernd Malter who performed numerous listening experiments as well as Carol Hill, Jeanne de Simon and Elviira Hartikainen who assisted in perfecting the manuscript.

Troy, NY, April 2005   *Ute Jekosch*

Table of Contents

1. Introduction 1

2. Aims and Methods of Speech Quality Assessment 5
  2.1 Summary 9

3. Aspects of Quality: Laying the Foundations 11
  3.1 Quality as the result of tests and assessments 11
  3.2 Quality as the design goal 16
  3.3 Designing the quality of innovative entities 18
  3.4 Summary 20

4. Speech Technology and Speech Quality Perception 23
  4.1 The priorities of modern speech technology 23
  4.2 Speech synthesis and its different design aims 28
      4.2.1 The history of speech synthesis: Emphases and aims 29
      4.2.2 Experience of techniques and expectations of quality 33
  4.3 Elements of quality of speech synthesizers 35
      4.3.1 The system input 35
  4.4 Text-to-speech synthesis: The text as an element of quality ...37
  4.5 Synthetic speech as a causal sign 41
  4.6 Functional internal system entities as elements of quality 46
  4.7 Speech synthesis and associated goals in research 47
  4.8 Summary 49

5. From Speech Perception to Assessment of Quality 53
  5.1 Perception and cognition 53
  5.2 Perception between individuals: Form and adaptation 55
  5.3 Inter-individual perception 58
  5.4 Speech quality assessment as a measuring process 59
  5.5 Measuring and measurands 61
  5.6 Measuring processes vs. investigative processes 64
  5.7 Measuring instruments and measuring organs 65
      5.7.1 Instrumental measuring methods (a digression) 65
  5.8 Summary 68

6. Quality Assessment in View of System Theory 71

7. Auditory Measuring Procedures 75
   7.1 Measuring and scaling methods in psycho-acoustics 75
   7.2 Measuring scales 80
      7.2.1 Standard scales 80
      7.2.2 The scale as a function of the measurand 82
   7.3 Summary 86

8. Formal aspects of speech quality measurements 89
   8.1 The measuring object »speech« 89
   8.2 Toward the quality of speech quality measurements 92
   8.3 Speech quality tests as measuring tools 96
      8.3.1 Intelligibility tests in speech technology and audiology 97
   8.4 Summary 102

9. Towards the Structure of Speech Quality Measurements 105
   9.1 Taxonomy for quality of speech synthesis 105
   9.2 Aspects specific to usage 107
   9.3 Systemic view of speech quality measurements 109

10. Segmental Intelligibility: A Dimension of Quality 113
   10.1 Intelligibility and comprehensibility tests: An introduction 114
      10.1.1 The Rhyme Test by Sotscheck 114
      10.1.2 The »SAM Segmental Test« 116
      10.1.3 The Semantically Unpredictable Sentences SUS Test 117
      10.1.4 The CLID Test 118
      10.1.5 General information on the CLID methodology 119
      10.1.6 Test vocabulary as a measuring influence 124
   10.2 On the comparability of study results 129
      10.2.1 On the influence of scaling 131
      10.2.2 On the influence of the stimulus context 133
      10.2.3 On the influence of syllable structure and frequency 134
      10.2.4 On the validity of the test results 139
   10.3 Summary 141

11. The Cluster Similarity Study 143
   11.1 On the background of the study 144
   11.2 Details of the study 148
   11.3 On rating and scaling 149
   11.4 Natural voice: Similarity profile of pre-nuclear consonant clusters 153

11.5 Comparison of the similarity profile of natural and synthetic speech 158
11.6 Similarity profile for various vowel contexts 164
11.7 Summary 165

12. Conclusion 175

13. References..179

Definitions 199
Subject Index 201
Author Index 205

1 Introduction

Voice and speech quality perception is a comparably young branch of mainly psycho-acoustics, metrology and linguistics. It deals with all processes that are connected with assessing the auditory quality of voice and speech. Isolated questions within voice and speech quality assessment have of course been dealt with for long mainly in phonetics, linguistics and related fields such as psychology and the cognitive sciences. However, these disciplines only had a limited interest in investigating and understanding the preconditions and processes of voice and speech *quality* perception itself. They were far more interested in using the results of voice and speech assessments to enhance their knowledge of the development and function of voice, speech and language and/or to increase their understanding of a human being's communicative ability. Voice and speech quality were therefore primarily a means to an end, not the purpose itself.

As part of the steady progress being made in the engineering context of information and telecommunication techniques, mainly speech quality assessment has gained in importance over the last years, particularly in speech technology. Even in this case the main interest was, and remains, to employ speech quality assessment as a means of releasing a further aim: Speech technology engineers require results from voice and speech quality assessments and measurements so that they can make certain statements on the product quality they are aiming to achieve in their individual R&D projects. Speech quality assessments are expected to fulfil certain requirements – they must make reliable and meaningful data available as quickly and cheaply as possible. This means that voice and speech quality assessment is again only an instrument, this time for engineers: Its results should demonstrate the quality of basic technical components of communication, they should denote characteristic qualities and quantities of speech technology systems, or for technological systems in general which are »enabling technologies« in communication technology.

Voice and speech quality assessment was unable to fulfil both tasks for a long time, mainly because the demands made were far in excess of its basic scientific research capabilities. This meant that speech technological products such as speech machines were already available long before suitable methods had been found to reliably assess and measure the quality of these techniques. In order for information and communication engineers to make their systems more acceptable, they desperately needed to know which components and features had to be improved in these systems' research and development phases.

The result was that speech technology engineers first developed their own methods to assess and measure speech quality. The main supporting science was psychoacoustics. These procedures, often conceived for very specific purposes, were later often taken as the unofficial »norm«, and – because there was a lack of suitably differentiated methods – then also applied in totally new contexts. Methods and procedures for assessing speech quality were used in new contexts, although their validity had not always been thoroughly checked beforehand. As far too little was known about the assessment logic, and as the influential factors of the assessment event could not be suitably identified and checked, voice and speech quality assessment was, with hindsight, often little more than a rough orientation which did not really measure voice and speech quality, but only roughly indicated the quality of some basic aspects of technological components.

Of course this inevitably led to dissatisfaction whenever statements on voice and speech quality assessment results had repercussions on decisions in engineering contexts, especially for competing cases. In the case of system design, displeasure could be felt whenever they had an important influence on the specifications, developments and production processes. An area of conflict developed between what it was hoped the speech-related psycho-acoustics would achieve and what it was, in fact, capable of. It slowly became clear at which points suitable methods and supporting knowledge were missing, and where simplifications and generalizations had been made that had led to a false picture of voice and speech objects being researched. The result of these dilemmas was, at best, a rather more cautious approach to speech-related psycho-acoustics, particularly when it was a case of dealing with quality experience.

In the light of these considerations there are many reasons why we should deal with the topic. This work will mainly concentrate on those aspects pertinent to research that are of particular relevance to practical applications in engineering contexts.

Traditionally, voice and speech are closely related to linguistics. The fact that linguistics has so far lagged behind in speech system research and engineering has led to important questions either not being posed by linguistics at all, or only being answered superficially. This becomes clear when we examine two questions concerned with the structure of tests designed to assess speech quality as an example. Their subject is the semiotic process, in other words processing speech as a carrier of information, as a linguistic sign [133]. The first question deals with the linguistic test material, while the second looks at scales:

Question 1: According to which criteria does the speech material have to be determined and selected when it is to be assessed in a laboratory, i.e. in a constructed and controlled listening situation? How does it have to be presented, if as a prototype it is to have all the important characteristic features of speech in such a way that it will lead to a corresponding assessment in natural or uncontrolled reception contexts?

Question 2: How is the assessment process to be initiated in a laboratory, and how will the resulting findings be collected? Which questions should be asked and which format, e.g. scales should be chosen so that each subject's response fits into it, thereby not violating to communicate prominent features each individual subject perceived?

The first question refers to the ambiguity of spoken test material: Whoever wants to carry out tests on voice and speech quality must first determine the speech test material, in other words, the test stimuli. Improperly, this test material is often regarded as a pure medium whose form characteristics the listener should judge. In view of semiotic theory speech is a carrier of information. As such it can be approached in a triadic relationship as an object of form, content and function. Still today, in many speech quality assessments, speech is nothing more than an object of form. However, the speech test material is always a sign carrier itself in the sense of a triadic relationship between from–content–function. If a spoken expression is improperly reduced to a static form object, it is highly unlikely that the stimulus will trigger off an equivalent or at least similar meaning for every test person. As known, it is the object of perception which is to be judged. This object is the result of an individual semiosis in which meaning is assigned to the perceptual event. Consequently, it is required that form, content and function of the stimulus test material are carefully selected. If this requirement is not fulfilled, it means that in terms of measurements the object has been defined (i.e. the speech sound as the measuring object), but those inherent quantities influencing the semiosis are not controlled at all. Therefore it is impossible to talk of measurements until this question has been answered. The validity of such assessments is questionable.

The assessment process has a number of coincidental factors that must be eliminated. Semiotics provides a way of doing this effectively: The stimulus is not regarded as an object that can be classified on a physical level, but is understood as an entity that is processed as a sign carrier. A sign is a relational and functional object of perception. This means that the choice of stimulus must be based not on the perspective of a static system, but on the perspective of a process, the semiosis.

The task is: If the relational or functional aspect of speech as a sign carrier is to be considered, how can it be chosen or structured, so that equivalent meaning is created in each test person's mind? The answer to this question is preceded by the creation of description categories which exceed determining speech stimuli as objects of form alone.

The second question belongs to the realms of semiotic theory as well. It can ultimately be put down to the classic de Saussure dichotomy of the signifier and the signified [223]. In this context it is necessary to examine which means can be used to communicate the thought of a judgment that is not due to an individually contextualized interpretation of speech, but predetermined by a pattern or format, e.g. a scale. Can an assessment of perceived speech be initiated and formulated in such a way that by standardizing and reducing its elements to the most important carriers of content a glut of information can be avoided, without detracting from the essential facts perceived or failing altogether to define such a format? The prerequisite for this is that the assessment of what has been perceived has to be succinct both on an inter- and intra-individual level. How can this condition be checked, bearing in mind that the assessment relates to a relationship and not to an object?

These introductory remarks should suffice to illustrate the necessity of analyzing the present state of speech-related psycho-acoustics from an interdisciplinary perspective and to relate psycho-acoustics to apparently heterogeneous scientific disciplines.

As aspects of different fields are brought together to give a global understanding of assessment processes, it is sometimes necessary to simplify facts by presenting them in the form of logical models. Logical models are always helpful when we are reporting on a number of specialist areas with each of them again concentrating on one single object of the numerous interwoven ones we are examining in the global clarification process. In this case the logical model is an abstraction of individual questions, forms a structure of elements and combination rules and has a definite, but limited analogy to the whole subject matter [168]. It is such a model approach by which many processes have to be simplified in this book. However, using models to support the facts is only then admissible when opposing theories describing and explaining the facts do not contradict the idea behind the model.

2 Aims and Methods of Speech Quality Assessment

Voice and speech quality assessment constitutes the main part of this work. It deals with the question of what listeners perceive from the mass of all the acoustic sound signals that are offered to their auditory system, which ones they select, which ones they cognitively process and how they assess them qualitatively and quantitatively. This process leads to an assessment of the perceived and experienced speech quality, addressing the following questions.

- What is quality?
- What demands are made on quality?
- What exactly is quality in the context of speech communication?
- What is generally voice and speech quality?
- Which aspects of the quality of speech can be identified?
- How can the quality of natural speech be defined?
- According to which criteria is speech quality assessed in specific contexts?
- Can the quality of speech be measured auditorily?
- Is speech quality always assessed in the same way?
- Which influential factors is the assessment dependent on?
- Can the quality of voice and speech be measured instrumentally?
- How can voice and speech quality be designed?
- What is the quality of artificial speech like?
- Which expectations are placed on the composition of artificial voices and speech?
- What handicaps are there when creating voice and speech artificially?

Relating to the field of speech technology:

- What does quality generally mean within the field of speech and communication technology?
- Are there conditions according to which the quality of speech technology systems should be assessed (speech quality versus system quality)?
- What demands in terms of quality are made of information and communication technologies?

These questions define the domain of auditory voice and speech quality assessment. Although the term »speech quality« generally has to do with grammatical and syntactical, lexically-semantic and semantically-pragmatic aspects, these aspects of quality, similar to written speech, will be largely ignored as measurands. Nevertheless, they will be taken into consideration as modifying factors.

In general the process of assessment presupposes events of every conceivable form of perception. This includes the reflection on what has been perceived, reflection on what is expected and an assessment of what has been perceived in terms of what was expected. Therefore speech quality manifests itself when the speech that has been perceived is compared with what was expected and is checked and assessed for the degree of its suitability. This process and its result will be referred to from now on as the speech quality event [18]. The term »speech quality« is defined as follows:

»speech quality«
> The result of assessing all the recognized and nameable features and feature values of a speech sample under examination, in terms of its suitability to fulfil the expectations of all the recognized and nameable features and feature values of individual expectations and/or social demands and/or demands.

Speech quality assessment is not a purely signal-controlled, deterministic event that will necessarily lead to the same conclusions whenever the same acoustic and environmental conditions are present. A speech quality assessment that relates to the same acoustic speech signals can produce totally different results when the process is repeated. There are various reasons for this. Consequently, the prerequisites, conditions and general processes of perception performance and the way a judgment is reached have to be analyzed if a speech quality judgment is to be reliable and meaningful after all. If the processes leading to a judgment are not understood, the judgments on speech quality will be questionable.

Understanding the assessment processes alone is not enough to provide a sufficient basis for understanding voice and speech quality assessment itself. For example, when undertaking the concrete task of assessing, you have to be aware of what has been left out. In order to be able to analyze the causality of some of the processes and to research the background of quality assessment, it is necessary to analyze the processes of speech perception and assessment under the following aspects:

- How do human listeners perceive features of voice and speech quality and how do they judge on them (procedures)?
- Which voice and speech quality features do they actually perceive (object)?
- Why do they judge on these and not on others (explanation)?
- How far-reaching is their ability to judge (scope)?
- How sure and how telling are their judgments (reliability and validity)?
- Which aspects are their judgments based upon (reason)?
- Are their judgments prototypical (representativity)?

The aspects mentioned here mainly apply to the assessment process and the degree of certainty of the judgment made. They are structural aids that describe events that have either taken place or are in the process of taking place.

Although voice and speech quality assessments are based on the experience and knowledge that arise from describing these events, there is in fact far more to it. One important aim in this field is to obtain findings on how voice and speech quality assessment can be introduced in a controlled artificial way. A basic requirement there is that the judgments obtained in this way do not lose meaningfulness when compared to randomly achieved judgments in natural environments. Another accompanying criterion is that the amount of effort involved should be minimized, e.g. with regard to the speech material to be assessed and the number of judging listeners.

The objective of voice and speech quality assessment as a scientific field therefore is to provide the knowledge to describe, structure, construct and execute processes that are essential to quality assessment. Very varied methods to do this are used: Apart from observing, describing and analyzing natural events, one effective method is to deliberately manipulate identified or suspected factors of influence. In this way assessment scenarios can be constructed in clearly defined and reproducible circumstances in a laboratory to increase knowledge of components and links.

Examining influential factors by means of manipulation is a commonly used method that will often reoccur in this piece of work. The focus of such experiments lies primarily not in examining the components and their relationships per se, but concentrates on detecting variables and constants in order to obtain a projection model of voice and speech quality assessment. Using a model to describe what has happened is necessary because speech quality assessment processes are very complex. Modeling is possible because, in spite of everything, the processes lead to similar results when similar circumstances are present and several listeners are questioned. Therefore the listeners' behavior cannot be put down to arbitrariness, rather it is strategic and subject to a certain systematology. The aim of this work is to analyze this systematology by means of a structuralistic approach and to obtain a simplified simulation, i.e. a projection model, which agrees with findings from other disciplines.

The principle used for the modeling assumes that the object, i.e. voice and speech quality perception, is a structured whole that can be described as a system. In this system the significance of the components that constitute the whole are characterized by a structure. This structure is an abstract model which is not subject to variations.

One such model is dealt with in more detail here. Once finished, it can be used as the basis of a design of artificial scenarios for quality assessment. A model-based design has the advantage that the artificial scenario does not have to have a reproductive character, in other words it must not be an extract of the world in which speech quality assessment naturally occurs, i.e. facts and processes must not be imitated or reproduced. The link between the natural and model-based assessment scenario is the analogous categorical structure. It should be pointed out that the analogy is based on the categorical structure, which does not mean that natural and artificial assessment scenarios are structured isomorphically. They cannot be structured iso-morphically because in each of the scenarios in which speech quality is experienced speech has a different communicative function.

In this book the field of voice and speech quality assessment will first be examined as a whole structural concept accordingly. In this concept structure is understood as a network of relationships. Therefore the field will be divided into components and their interconnecting relationships. With regard to an integral structural concept of voice and speech quality assessment, those components that have already been mentioned are created and ordered in such a way that any change to an individual component will result in a change to all the others that are associated with this one. With the help of such a projection model, the whole domain of voice and speech quality assessment is to be understood as a system.

If certain components are viewed individually, with no reflection on the context of speech quality assessment as a whole, they prove to be central objects of individual areas of knowledge, e.g. of psycho-acoustics, epistemology, metrology and linguistics. In each of these areas one or more of these components that are involved in speech quality assessment are treated as individual complex subjects of research. Due to the variety of questions and facts thrown up by each individual science in the context of this work, we can only examine these components under systematic aspects and concentrate on those aspects that directly contribute to the comprehension of the processes in voice and speech quality assessment. Therefore the simulated projection models that will be developed here will reproduce associated aspects as exactly as possible to avoid contradicting individual sciences and will be as simplified as possible to promote voice and speech quality assessment. This book will therefore not rise to the challenge of reviewing in detail every aspect from the point of view of each science. Generalizations will be made where they can contribute to the integral structural analysis of voice and speech quality assessment.

At first sight there appears to be a contradiction between the method chosen, i.e. the structural description, and the domain under investigation, i.e. voice and speech quality assessment. The basic idea behind the structure initially implies something static, while voice and speech quality assessments are dynamic events. Within the framework of a structural description components are certainly named, but their materialism fades into the background just as much as the movements of, changes to and development of the structured whole. Kröber describes structure as "relatively constant", as "curdling dynamics", and as "the cross-section through development to a particular point in time $= t_x$" [156].

Now although neither the production of acoustic speech signals nor their perception and assessment are static creations, this is the approach generally used. The above-mentioned doubts are not justifiable to such an extent that they can afford to ignore the modeling capacity as a major characteristic of the model or the aim of the modeling experiment itself. A projection model remains, from necessity, a fairly rough model. It is definitely insufficient as the sole basis of an algorithm for speech quality assessment. Therefore the ideal model is one that best suits the described set of circumstances and possibilities of voice and speech quality assessment. The domain has a dynamic character, but the dynamic element is neither a whim nor infinite. The projection model attempts to analyze this dynamism systematically. Thus speech quality assessment is understood as a time-variant system.

2.1 Summary

The aim is to systematically analyze voice and speech quality assessment by means of a projection model, whereby this model consists of the components and relationships between the individual components. This produces a dynamic network which can analyze variants in time and position. The projection model is designed in such a way that no known facts are omitted.

Having reached this goal, the model itself becomes a means of identifying gaps in our knowledge and formulating basic questions concerning speech quality assessment. The advantage is that these unanswered questions can be accommodated in the structural whole and that its content is a result of the questions' relevance to the structural whole. In addition the model is an aid for conceptualizing a data survey on speech quality experience.

3 Aspects of Quality: Laying the Foundations

Whenever we hear the term »quality« we frequently think of the quality of goods and products: They can be of good or first class quality, durable or proven quality. In everyday usage the term» quality« principally has a positive connotation. Quality is generally a good characteristic. If an item is bad, it has no or insufficient quality. The persons who pass judgment on quality are the users or the customers. Whatever they find useful has quality.

The term »quality« can be found in different contexts. Amongst other things, it plays a decisive role in the field of industrial production. Here the term has been clearly defined. Quality management, design and project control, tests, corrections etc. are all facets of quality, which within industrial production serve to eliminate production errors before they can occur, thus increasing productivity and competitiveness.

Before discussing the tasks and possible solutions within voice and speech quality research, it is essential to clarify some of the terms involved. One reason for this is that the term »quality« also occurs in some derivatives such as quality checks, elements of quality, quality features, requirements of quality, planning of quality, product quality, test quality, and/or quality of test implementation. In order to be able to inaugurate these nuances into the field of voice and speech quality assessment, these terms must be given a standard definition. As will later become clear, many of the aspects of the term »quality«, as they are used in the field of industrial production, are also applicable to voice and speech quality. Therefore a global definition of quality will first be derived and then modified to specifically apply to voice and speech quality assessments. Well-known definitions taken from various fields will be included, primarily from the areas of quality assurance and statistics that apply to industrial production.

3.1 Quality as the result of tests and assessments

The starting point for our debate on the term »quality« is the following definition:

DIN Def. »quality«
> "[...] composition of an entity with regards to its ability to fulfil determined and necessary requirements." [49]

DIN Def. »composition«
> "[...] the totality of the features and the values of the features [...] of an entity [...]." [49]

DIN Def. »feature«

"[...] characteristic for recognizing or distinguishing between entities [...]."
[50]

DIN Def. »value of a feature«

"[...] the value attributed to the manifestation of a feature." [50]

This norm is specified by the following comment:

"By specifically determining the feature in question the type of feature (e.g.
color, length) is measured [...] and thus the type of feature value (e.g. red,
3 m)." [49]

DIN Def. »entity«

"[...] material or immaterial object under observation." [49]

These definitions form a suitable basis for determining the terms of voice and
speech quality, but leave some questions unanswered. For example the definition of
the term »feature« describes how features function (they are for recognizing and differentiating between entities). However, the term conveys information neither on
the process of determining the contents nor on the contents themselves: Characteristics exist that are for recognizing or distinguishing purposes only. The following
questions, which will later prove extremely important for voice and speech quality
research, yet remain unanswered:

- Who chooses these features?
- How are they described?
- Can general features be identified for one and the same item, irrespective of the
  individual observer or user?
- How high is the degree of accuracy when the features are described in terms of
  their agreement with a set of features that is, in theory, objective?
- How can it be ensured that all the typical features have actually been listed?

Just as the definition of feature leaves much to be desired, the definition of quality
raises a few questions: The entity itself must fulfil predetermined requirements to
realize quality. But:

- Who decides which features of an entity are relevant to quality?
- Who specifies the norm to which they are compared?
- According to which features will a test have to be carried out to prove agreement
  between the composed entity and predetermined necessary requirements laid
  down in norms?
- Is the choice of quality features based on facts and can it be justified?

Depending on the object that is to be assessed in terms of its quality, these questions
can be interpreted differently. For example in industrial manufacturing quality features of technical end products destined for the general public are clearly specified.
They mainly refer to aspects of consumer functionalism and safety of products and

processes, e.g. motor vehicle inspection. In such cases they are logically expressed as mathematical formulas, metricated and formalized. Technical products are generally described in terms of their functional, technical and ergonomic characteristics, whereby this mode of description relies heavily on normed formalizations. There are often catalogs of standard requirements specific to each product which serve as the expected norm and are considered obligatory when quality assessments have to be made. Data on quality (here in the sense of test results) are attained by comparing how the product performs and which features are required in the name of functionalism. In this context the term »function« is used in the sense of processed characteristics.

Provided that the quality of the function of technical products is to be examined, a list of characteristics of the typical functional features is an essential aid. The observer can concentrate on following, for example, the course of physical events without having to describe the characteristics of single events in any detail. In this case the term »quality« is defined as follows:

ISO Def. »quality«
"[…] totality of characteristics of an entity […] that bear on its ability to satisfy stated and implied needs." [52]

Apart from the fact that this definition also omits to answer the question as to how "stated and implied needs" are determined this definition is deficient in the following way: What happens when the quality of the »form« of an entity, e.g. a work of art or speech has to be determined? The perception of a functional object is based on features that have a logical coherence. Structural features can be comprehended as the physical interrelation of processes influenced by the presence of existing features. Each individual may perceive a form differently. For the time being satisfying systems for capturing form related features of voice and speech quality do not exist.

But even if such systems existed, another problem would come to the fore: It is directed towards the relation of externally controlled and externally uncontrolled perception, seen from the point of view of the perceiving subject. If a standardized system for describing features would be available for laboratory tests, how can it be guaranteed that the perception performance of form objects in an arranged and controlled artificial test environment occurs in the same way as if perception had been fortuitous or random in a natural situation? Deficits in the awareness process may occur in laboratory environment in the sense that there are features which are potentially perceptible but not in fact perceived because of guidance. This shows that the significance of a standardized system of form characteristics of the entities to be assessed limit the meaningfulness of quality assessments considerably. In view of a systematic approach to voice and speech quality assessment the following differentiations have to be made between »controlled perception« and »random perception« as well as between »controlled quality assessment« and »random quality assessment«:

»controlled perception«

Result of reflection on an externally initiated (extrinsic) perception event that has been artificially evoked.

»random perception«

Result of reflection on a casual (intrinsic) perception event that has been evoked naturally.

»controlled quality assessment«

The extrinsically initiated process of comparing the entirety of the perceived features of an observed, representative entity in terms of its suitability to fulfil all the individual's and/or group's expectations of features. The result of this comparative process is the quality assessment.

»random quality assessment«

The intrinsically initiated process of comparing the totality of the perceived features of an observed, representative entity in terms of its suitability to fulfil all the individual's or group's expectations of features. The result of this comparative process is the quality assessment.

If there are no predefined features guiding perception and assessment, the quality of objects is assessed anonymously, a process that should not be underestimated. This is particularly true for the quality assessment of the form of speech. Listeners in natural communication situations can decide relatively easily and spontaneously when they have perceived sufficient or insufficient quality of the spoken form, but they often have difficulty in naming and describing in detail exactly which form features determining quality dominated speech quality perception. Moreover, they invariably have problems in giving differentiated reasons and communicating their assessment of speech quality. It is irrelevant whether listeners passing judgments are acquainted with the speech they are to assess, whether they have heard it by chance, or whether it appears to them unreal or strange. Difficulties in naming quality features obviously occur because voice and speech quality has no contours and is unclear in a complex communication process. There are rarely quality features that are clearly distinct from each other, giving the quality event a clear framework.

The same is true for naming desirable and perceived features of speech forms. The quality of the form of speech is taken for granted and therefore seldom questioned. The difficulties mentioned above increase when the listener is confronted with unexpected or unknown speech signals. Synthetic speech is a good example of this, at least for first-time listeners. In order to be able to analyze the individuality of random perception performances of forms, it is important to revise the definition of the term »feature« according to [50] for the field of voice and speech quality assessment as follows:

»feature«

A recognizable and nameable characteristic of an entity.

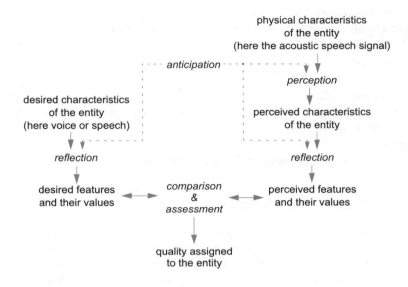

Fig. 3.1. A schematic description of the quality event

This definition contains a further aspect which is important for the term »feature«: A feature is not only a recognized characteristic, but this characteristic is also communicable. A feature serves to describe and/or contrast entities. By naming the typical features and properties of an entity or its sub-entity (entities), the special features that form the basis of its distinctiveness from other entity or sub-entities come to the fore.

If observers want to assess the quality of an entity they compare the features of the entity they perceive (data of perception) with the required features (data of expectation) [133]. Only if the actual features of the perceived entity live up to their expectations of quality, will the entity be deemed suitable and thus of quality. It is therefore necessary to analyze those features and their respective values that have proved to be effective in a natural speech perception process not only for its desired but also for its perceived composition (cf. Fig. 3.1.).

From this the following definitions can be deduced. In this form they are binding for the rest of this volume.

Working Definitions:

»quality«
　　Result of judgment of the perceived composition of an entity with respect to its desired composition.

»perceived composition«
> Totality of features of an entity. Signal for the identity of the entity visible to the perceiver.

»entity«
> "Material or immaterial object under observation". [49]

»desired composition«
> Totality of features of individual expectations and/or relevant demands and/ or social requirements.

»feature«
> Recognizable and nameable characteristic of an entity.

3.2 Quality as the design goal

In the preceding sections quality was primarily discussed from the perspective of interactive users of entities, who assessed entities in terms of perceived quality. However, the results of such assessments can be used by different groups representing various interests as a means to achieving each group's specific aims. Developers and producers who want to create high-quality entities are typical examples of such interest groups. Their aim is to create entities in such a way that the user attributes the highest possible quality standard to the entity. The success or failure of this aim is ultimately dependent on the developer's ability to predict the typical user's behavior in relation to quality assessment: The more comprehensive the investigation into the user's expectations of the composition of the entities is, the stronger the foundation for the entities' design, and the greater the probability is that the user will ultimately attribute quality to the entities. For the different utilities cf. [132].

In order to produce high quality entities, components that make important contributions to create quality still have to be identified, classified and evaluated in the individual phases of the so-called quality loop. According to [49] the quality loop covers the planning phase (market research, conceptualization, design, testing, production planning), the realization phase (procuring, production, final testing, storage, distribution) and the usage phase (maintenance, disposal).

Such components the function of which can be seen as a contribution to the quality forming process are called elements of quality. The analogy to this is the quality feature, where the perceived form of the feature of the created entity is understood to be functional in terms of the quality assessment.

Accordingly the terms »element of quality« [49] and »quality feature« are defined as follows:

DIN Def. »element of quality«
> "Contribution to the quality
> – of a material or immaterial product as the result of an action/activity or a process in one of the planning, execution or usage phases

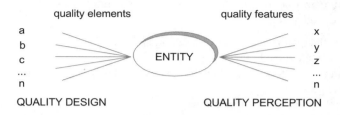

quality elements   quality features

a   x
b   y
c   ENTITY   z
...   ...
n   n

QUALITY DESIGN   QUALITY PERCEPTION

Fig. 3.2. Quality elements and quality features

– of an action or of a process as the result of an element in the course of this action or process."

Def. »quality feature«

A quality feature is a recognized and designated characteristic of an entity that is relevant to the entity's quality.

While an element of quality is the building block for designing an entity, a quality feature is the analyzed result of the perceived, designed entity and is therefore the basis of any description of its quality. According to [50] neither a single quality feature standing alone nor its values can be described as quality or an element of quality. It is therefore important to ensure that the term »quality feature« is not simply seen as a direct characteristic of an element of quality: An element of quality is a component of the quality-creating phase, e.g. in the design or production phase.

According to the definition the quality of an entity is the result of the degree of agreement attained between the desired and perceived composition. If the highest possible degree of agreement is achieved, this is considered to be a successful creation of quality.

»successful quality design«

Elements make their mark in the quality shaping process of an entity in that their finished forms are offered to persons who perceive and assess quality. They assign quality features to the features of the entity, or its perceived sub-entities, that they deem to be identical to those they regard, in the sense of individual expectations and/or social demands, as desirable or positively expected in the entity and its sub-entities.

Our perception of quality features (and only indirectly the structure of the elements of quality) ultimately determines the quality of the entities. But what quality must the individual elements have so that they contribute as much as possible to the quality of the total entity, and thus constitute the best possible elements of quality?

It would be both simplistic and inappropriate to see every element of quality as an independent entity that has to have the highest possible quality if a top quality host entity is to be created from all the elements of quality put together. These elements should only be sought when they contribute to the optimum functional quality in the target entity.

- What are the quality features of voice and speech?
- What characteristics do they have?
- How can they be transformed into elements of quality in the design phase e.g. of artificial speech?
- How do these elements interact?
- How are certain artificial speech sounds of the elements of quality perceived and assessed by the listener?

Take the task to synthesize speech as an example: To avoid the created entity becoming a product of chance, and to be able to make decisions on its components, you have to know what composition is desired, how this can be adequately described by different groups of users in terms of its features, and how generalized individual and/or group expectations can ultimately be. Speech quality assessment serves to supply useful information, not only at the point in time when specifications and plans are formulated, but throughout the whole system development phase.

3.3 Designing the quality of innovative entities

The demands made of quality design increase when quality is understood as the degree of correspondence between what is at present available and what is at present imaginable. Reitzle even goes a step further: "The optimum available quality of a product or an activity is considered to have been achieved when the customer's expectations have been exceeded." [215]

To avoid quality design becoming a process dominated by pure speculation, certain prerequisites must be given and certain conditions met. This can be illustrated using an example taken from speech-related communication techniques in virtual realities.

Although today's technology can already provide us with a multitude of appliances and facilities that are continually widening the natural field of speech acts, information and communication technologies continue to strive for improvement and new inventions. They are therefore partly responsible for creating, promoting and reinforcing social demands.

Recently it has become clear that the modern research and development goals of these technologies are no longer directed towards reality. Present and future technologies are concentrating more and more on the virtual plane. This field is also known as phenotechniques. The aim of this scientific field is to simulate the perceived world, to create and store new worlds so that they exist independently of real space and time (cf. [222]).

Today the vision of letting oneself be transported to an artificial world and being able to communicate in it can no longer be dismissed as pure fantasy. For example, there do exist multi-sensory simulation techniques used for airplanes. Here trainee pilots can be transported to another sensory environment which they have never experienced before.

It is definitely feasible that the human multi-sensory, cognitive system could be guided by simulation techniques in such a way that the perceivers feel that they have entered another world (so-called immersion) [210]. Today the thought of meeting someone at a virtual place for a meeting without having to leave the office, but still being able to communicate visually and auditorily no longer appears to be unrealistic. It is even feasible to be able to experience the virtual world through touch and smell. The long-term goal of information and communication technologies was to research facts, to explain, present and change things that already exist in such a way that our thirst for knowledge was stilled. Maximum profits and social purposes were also main targets. Today these experts are striving to create the basis for new facts, so that temporal and spatial structures can be overcome [184].

Structuring and exploring this field can be programmed to include future activities. This enables communication technology to gradually forsake its traditional ground and to take up its new position facing the challenges of designing a vision of intellectual reality. In doing so technology is striving to extend the contents of natural perception and intelligence by means of phenotechniques and electronic information processing.

Of course this assumes that it is known what reliable conditions actually are. In this case perceivers can really forget the material basis they know and trust and can be guided to points of reference for forthcoming activities on which their perception will then be based. However, even if this were technically possible, a number of questions concerning the user would remain:

• What is the desired and actual reaction of the user to phenotechniques?
• How well can we expect innovative technologies to be accepted?
• Which prerequisites have to be fulfilled before such systems can be integrated into daily life?
• Which conditions do the users really expect tele-virtual systems or their components to fulfil?
• What expectations do they have?
• Do they have any expectations at all of such a seemingly strange application?

These questions have to be answered against the background that those kind of systems should be developed which are acceptable to the users. Quality has to be designed. Quality has been defined as the result of an assessment of the perceived composition of an entity with regard to expected attitudes and demands. However, the expectations and demands of future users of innovative technologies remain a mystery. At the point in time when the innovative entity is designed, the specifications of the expected composition are based on analogies, or are purely speculative. The more innovative and exotic the entities are, the more speculative the prediction of user behavior is.

Predicting future user behavior will – within limits – always be speculative, because the constancy of attitudes towards expectation generally depends on the tempo of the social change. The faster society changes, the faster expectation and demands of entities change. If expectation and demands change the quality attributed to the entity is also forced to change, even if the same physical composition of the entity is the basis of the user's personal assessment. This assessment changes because the user's perspective has changed. Innovative system developers have to carefully analyze this dynamic structure and take it into consideration whenever they make decisions.

However, the opposite can also be the case: The group of users targeted by the system developers can behave unexpectedly, reacting inflexibly to change and rejecting a new technology.

Although designing innovative entities is a risky business, the risk can be considerably limited and minimized. This can be achieved by first identifying the factors of perception and quality assessment as exactly as possible, and then analyzing their influence on the assessment process and the resulting judgment. In addition it must still be possible to predict the dynamic reciprocal effects of elements of quality in the design plans.

If this discourse is applied to voice and speech quality assessment it means that, given these circumstances, the aim of all those involved in carrying out research, developing and initiating voice and speech technology must be to recognize the highest attainable goal. This goal can be very concrete, but it could equally be speculative. The further it sinks into the realms of speculation, the greater the onus is on science and research to recognize and describe the basic conditions, for example using models, to enable the application to be put to good use.

3.4 Summary

In this chapter the term »quality« was first discussed in the context of industrial production in order to be able to take advantage of the fact that it can be associated with a known quantity. Different aspects and components were extracted which defined the framework of a global understanding of the term »quality«. It was not attempted to profess that the contributions on the term »quality« are complete or could not be differentiated further. No effort was made to illustrate all the many facets of the term »quality« – this was not the designated aim. The real aim was to throw light on the term »quality« from different perspectives and, in particular, to highlight that quality does not exist, but does evolve.

• Quality is not absolute, neither is it the property of an entity, something inherent the presence of which can be taken for granted. Basically quality is attributed to an entity by the individual who observes and assesses it.

• The quality event is controlled by expectation. As individual expectations can change, expectation of quality (the expected composition of an entity) is dependent on time and place. Quality is dynamic – it comes into being when it is combined with the desired and perceived composition of the corresponding entity in a spatial and temporal framework.

• The object to be assessed can be viewed from different perspectives. It is important to differentiate between the contents, function and form of the object.

• How the contents and functions of these objects are assessed has been laid down by a standardized checklist of norms in many cases. This specifies how the entity is to be viewed. If all the specified features are seen as a whole, they are sufficient to give comparable assessments on the quality of the entity which can be meaningful and inter-individual.

• Assessing the form of the objects can easily be carried out using existing systems for describing features. The developer of such descriptive systems has to avoid drawing attention to unrealistic issues to prevent an intolerable deviation between the controlled and natural behavior when making an assessment. Therefore it is important to distinguish between controlled and random perception, on the one hand, and between controlled and random assessment, on the other.

• There must be a clear differentiation between an element of quality and a quality feature. An element of quality is the foundation on which the design of an entity is based, whereas a quality feature is the result of having analyzed the perceived, designed entity and which leads to a description of quality.

• Quality can be viewed from the point of view of the person who perceives and assesses, whereby the aim is to reveal all sorts of behavior and basic principles. But it can also be analyzed from the perspective of system developers. Their aim is to design entities that are capable of being categorized as top quality by the person carrying out the assessment.

• When designing entities, the targets cover the whole spectrum between research and development. This must be taken into consideration in the assessment process: Therefore a clear distinction must be made whether the quality of an object of research (which is perhaps characterized by several questions for which there are no answers) is being assessed, or whether the assessment is of an object of development (which is only the result of a conversion).

4 Speech Technology and Speech Quality Perception

In this chapter speech technologies will be described from various perspectives: From the point of view of research and development, system use and quality experience, as well as quality measurement. This is not done to validate the relevance of the technology itself, but to authenticate the perspectives, demands, conditions, aims and limitations of voice and speech quality assessments by presenting facts. They establish a data-based approach to analytically review the status of speech quality measurements. Speech technology is introduced here as an instrument. The issue is not to give a state-of-the-art of speech technology but to explore in how far speech technology can be used to examine the fundamentals of voice and speech quality perception. In subsequent chapters different experiments are introduced where speech technology devices are the measurands; however, these experiments are a means themselves to exemplify which quality assessment processes can be observed and how they can be analyzed. It is not experimental data that are of interest primarily but processes of audition and judgment which lead to these data. In order to understand what these experiments are driving at it is useful to have a broad overview of voice and speech devices and their performance. Positioning speech technologies in their various contexts highlights the need to discuss voice and speech quality perception and to collect evidence that will justify the approach introduced here.

4.1 The priorities of modern speech technology

In general, speech technology follows the aim of supporting natural communicative possibilities involving people by means of technical devices and facilities. It strives to emulate the functions of naturally spoken communication, or to create new ones in order to be able to enhance human possibilities of communication. Among those functions copied are, for example, the production, transmission, recognition and processing of speech. Some of the basic aims and core areas of speech technologies are dealt with in the following.

• Speaker (sender): When using speech synthesis certain capabilities of the speaker are modeled. These capabilities include, for example, reading a text out loud, formulating news in speech from key information that can be found in the system's input, e.g. generating and reading aloud a weather forecast based on meteorological data [56], [74], [75].

• Channel: The aim of digital speech coding is to represent acoustic signals digitally and to make the size of an audio recording of speech smaller in order to facilitate faster transmission of speech over digital networks. Another application of speech coding is to create smaller audio files for efficient storage media. Finally, voice over internet protocol (VoIP) phone systems are critically reliant on effective coding and compression algorithms. Criteria of digital speech coding algorithms are speech quality, bit rate and coding/decoding effort required [74], [181], [180], [213], [161].

• Listener (receiver): Speaker identification and speech recognition are related to listener performances: Speaker identification aims to define characteristic minimal units that identify the person who is speaking. Algorithms for processing spoken utterances enable an unequivocal identification of the speaker. Areas of application include forensics or security checks for entry into restricted areas [200], [199].

Just like speaker identification, speech recognition uses acoustic speech signals at the system's entrance port. However, its aim is to recognize the semantic contents of a piece of speech in order to transform what has been said into another format (e.g. converting the utterance into a written form or a system action), even for subsequent automatic speech translation [221], [186].

Audiology and hearing aid technology are just two of the fields within speech technology that are concerned with the hearer's performance. They involve manipulating the speech sounds presented to the ear in such a way that those people who have hearing deficiencies can process the acoustic speech signals better. These technologies enable people who are hard of hearing to understand spoken utterances, despite their individual limitations. One aspect of this is to locate the speakers in a complex environment with noise interference, or in multi-speaker situations. Here results of binaural technology come into play in that they simulate the effects of directional hearing, thus supporting the division into useful and disturbing signals (the so-called cocktail-party effect) [24], [34]. Speech quality assessments are also carried out in this field to analyze the capabilities of the systems.

These examples demonstrate that many special technological contributions in different fields linked to spoken utterances have been made. Their real value becomes clear in practical contexts i.e. when the speaker and/or the listener decide whether a certain technology offered to them is useful or not.

If the quality of processes in the field of speech technology is to be examined, the first step is to clearly identify the aim and function of the quality analysis. In this context not exclusively the device alone but also the framework and conditions under which speech technology engineers work on special research and technology issues can also be direction-setting for speech quality assessment. First a distinction has to be made between a quality assessment that accompanies research and development projects, and one that is oriented towards the consumers and users:

• Should the technology be assessed from the point of view of researchers and/or engineers, who – amongst other things – rely on potential users providing feedback so that they can recognize possible research and development errors as soon as possible (quality assessment in research and development)?
• Should the technology be seen from the point of view of the end users who hope to get a supporting device for an improvement in the communication possibilities available to them (quality assessment oriented towards consumers or users)?

In this chapter quality assessments of speech technologies – preferably speech synthesis – will primarily be dealt with from the perspective of research and development, even if questions relevant to applications should also arise. Aspects that are relevant to quality are described in each phase of the »quality loop« (cf. page 16), beginning with the idea itself, via planning and specification to constructing the actual speech-technology prototype.

As it tended to accompany the research and development phase of speech technology engineering, voice and speech quality assessment was, and still is, usually given a minor role. In this context speech quality assessment's task is to analyze and provide only that information which speech technology engineers require at a specific point in time to realize their individual aims in researching or constructing a speech-technological entity. For speech quality assessment this means that the quality of their results are measured not by what is principally viable, but strictly by the experts' expectations. In this context speech quality assessment has to adhere to very strict guidelines, which means that the careful choice or even development of new appropriate methods and means have often to take second place.

The consequence of this is that, in general, existing knowledge of assessing speech quality cannot be exhausted. Compromises are often chosen and simplifications preferred. Therefore it is mostly essential to test technological components whose qualities are to be analyzed in a laboratory. The reason is obvious: What happens in a laboratory is always neutralized for speech communication, all the seemingly important influences can be controlled and the behavior of the subjects can be correlated with the speech sounds on offer. Compared to natural communication situations, the assessing process loses its anonymity and becomes more transparent. The following examples illustrate this more clearly:

If a quality assessment of a source and/or channel is to be carried out, people with normal hearing abilities are usually asked to give their assessment. If, however, as in the field of audiology or its related technologies, examining hearing disabilities is the main focus, the characteristics of the source and channel are mostly kept constant so that, for example, the patients' remaining hearing deficiencies can be analyzed.

A comprehensive speech quality assessment of a speech-technological prototype for an application covers every conceivable aspect of communication events. As an example, people with hearing disabilities can, of course, have a special interest in using a synthesis system, bearing in mind that synthetic speech will soon become available for certain services via wire-bound or the cellular network. But when discussing these aspects the onus of the discussion changes. The choice of paradigms and methods is always dependent on the defined aim. In research setting limits and concentrating on central parameters under laboratory conditions is essential for examining correlations between the parameters under controlled conditions. However, the following should not be forgotten: In terms of speech technology the data only reflect the listeners' reactions to elementary events.

Moreover, the same is true for the speech material, namely the test vocabulary: In order to specifically research causes, for instance, in speaker identification or speech recognition, it can be very interesting to analyze the system's performance using only disturbed or clear speech sounds, to have a system input comprising only the speech sounds of those with speaking disabilities, or to base all the experiments on the spoken utterances of average speakers. But even here the results are undeniably linked to the research aim. The paradigm for speech quality assessment is therefore predetermined by the aim of the experiment, and the aim is specified by the speech technology engineer. This proves that there is never only one single test for speech quality assessment, just as there is also no definitive quality correlated with speech technology.

The necessity of differentiating between the above can be demonstrated by the following examples. It is not true to say that there are only isolated cases where hearing impaired people who wear a hearing aid understand speech well in the framework of speech audiometry tests in clinical contexts, but have problems making themselves understood in their own private environments. When the hearing aid is fitted, the used speech material often originates from a cooperative speaker (in terms of hearing disabilities) and has little or no disturbing noises [47], [48], [97]. Sitting in a clinical laboratory, the persons with hearing aids are in an ideal situation to hear speech. In a more complex and more realistic communication situation they face a variety of disturbing factors. Tests available are limited to peripheral disturbances when processing speech [149].

The same is true for distorted or incomplete sound sources, e.g. for the quality of synthetic speech. If synthetic speech is tested in a laboratory the test results are often quite promising. But if synthetic speech is broadcast from the loudspeakers in a railway station forecourt, problems in understanding can easily be anticipated. If you imagine that such a synthetic voice contains information that a person with hearing disabilities wishes to listen to in an environment where there are disturbances, it becomes obvious that the meaningfulness of speech quality assessments is always linked to the conditions of the survey and cannot simply be transferred to other situations without changes being made. Or, to put it in semiotic terms, speech quality is functional.

Assessments of speech quality are therefore a result of the whole survey. In order to be able to understand these results properly, their position in contrast to other results has to be examined. Part of this involves answering the following questions:

- With which predetermined aims, under which preconditions and within what framework are speech technological components or devices developed?
- What are the real aims in system development?
- Are they clearly specified?
- Which aims of other interested parties are considered when the quality design process is initiated, and under which conditions does the process actually take place?
- Are the given aims the objective of research or development?
- Which methods and methodologies exist for speech quality assessment?
- What areas does each one cover?
- For which predetermined aims, under which conditions and within which framework were they developed?
- Have these methods been scientifically proven?

The last question points to the fact that speech quality assessment itself is, to a large extent, an object of scientific interest and must therefore constantly respond to new challenges. In its function supporting research, voice and speech quality assessment is sometimes confronted by new phenomena that require a thorough analysis. Assessing synthetic speech or acoustic speech signals in virtual environments are typical examples of this. Here the hearers are confronted with new facts. Not only for speech technology engineers but also for experts on voice and speech quality it is important to know how listeners react to unknown signals, what strategies they develop to process these signals, how their perception is steered by background knowledge – but also how other sensory perceptions influence quality experience. Although some of the basic knowledge is still missing, speech quality assessments in research and development are still expected to provide the basis for constructing test scenarios that come up with reliable results. The consequence is obvious: If the events correlated with the assessment process are not sufficiently researched and the influencing factors not clearly identified and classified, the results will be unreliable or even questionable.

This is made clear by the development of set aims in speech synthesis, with which speech quality assessment is not always directly connected. With regard to developing predetermined aims in speech synthesis techniques, the main focus has changed in the course of time: At the beginning the aim was to be able to create sounds that bore some similarity to speech. Today it is audio and video synthesis [225] or speech synthesis which, in combination with other technologies such as speech transmission, speech recognition, speaker identification and computer-based translation, is even capable of forming a dialog [182]. Moreover, speech synthesis should be adaptable to communication requirements, it should be able to imitate individual ways of speaking and/or to adapt to situations involving noise interference [153].

In technological terms much is possible today. However, we still do not know enough about voice and speech quality perception to control technological components, for example in voice generation. This means that at the moment the main task consists of moving away from the atomistic idea of spoken speech towards analyzing the many varieties of natural speech and its use in communication [161]. This is naturally a very ambitious and challenging aim, because it means copying by technical devices the flexibility of a natural partner in communication, including emotional aspects in speech [37]. To meet this objective, technologies must be infused with detailed knowledge. This is, in particular, a matter of speech and communication sciences.

One of the main tasks of voice and speech quality assessment is to act as a support and to analyze the links between acoustic speech signals and the listener's assessments and behavior. This means that synthetically produced signals are no longer just the object of research, but the means of the assessment. It is eminently suitable, as synthetic speech has an advantage over naturally spoken speech in that it is easier to manipulate and reproduce. As a consequence, analyzes with very different influential components can be undertaken so that different results can be compared with each other.

More detail on this topic is provided in subsequent chapters. The experiments described there are used to assess the quality of speech synthesis systems according to the standards set by science and research. At the same time they examine auditory behavior and assessments. The boundaries of these kinds of processes and the points where methods need to be optimized are also addressed. This provides the synthesis engineers with valuable results and data, and, at the same time, forms the basis for a detailed scientific debate on the topic.

4.2 Speech synthesis and its different design aims

With regard to speech synthesis, engineers aim to develop a system which, similar to humans, can carry information, make announcements, or read texts out loud. In historical terms the starting point for constructing a speech synthesis system was the atomistic idea of the essence of speech and the aim to copy this using technical means. However, it soon became apparent that natural speech sounds consist of far more than the concatenation of single speech sounds, syllables, or words. Speech does not comprise sounds that act as single events like letters of the alphabet which are put together to form words and sentences. Speech may consist of a tightly knit network of speech sounds, but additionally there is the matter of structure and meaning. Understanding speech always involves guessing and predicting words in a complex interaction between eyes and ears, memories, expectations, experiences and feelings. The listeners' brains mobilizes all their faculties to make sense of what their partners are saying. Hearing does not end with the ear. In the course of developing a speech synthesis system this fact was clearly audible.

4.2.1 The history of speech synthesis: Emphases and aims

In the infancy of speech synthesis the main interest was to copy man's ability to speak using technical means. The goal was to build a functioning model of the human speech organs in order to gain more knowledge on how human speech production worked.

It began by observing how speech is produced: Speakers put the information they wish to impart into a spatial code. They move articulatory organs involved in speaking into a certain position. In principle it is already possible to understand what is to be articulated in the form of this spatial code by scanning the positions of the individual organs of articulation and their changes as a function of time. This scanning means that there has to be a direct connection to the speaker's articulatory organs [108]. However, speaking means to detach the speaker from the listener, and this requires a more suitable signal. Air is pressed through the throat, mouth and nose tract whereby its intensity is regulated by the lungs. This produces variations in airflow, modulated by the organs of articulation, via which information can be passed over limited physical distances. If these variations in airflow are in the audio range (approximately 16 Hz to 16 kHz for people with normal hearing abilities), they can be perceived, analyzed and decoded as speech sounds by the listener's auditory organ.

There have been many attempts to construct a speech synthesis system. The very first step involved constructing a voice synthesizer. This is an entity of quality that can produce speech sounds independently of human articulatory organs. The first successful voice synthesizer was made by Kratzenstein at 1779 [155]. Kratzenstein constructed a variable acoustic resonator whose form was similar to the human vocal tract. His reason for undertaking such a project was a competition run by the Academy of Sciences in St. Petersburg. Whoever could artificially demonstrate by means of a model the auditory differences between five vowels would win the prize. Kratzenstein came first [5].

Kempelen's voice synthesizer of 1791 is far more renowned [54], [55]. He managed to produce 19 consonants using a pair of bellows, a pressure chamber, molded nostril-like openings and diverse switches. This construction can be seen in Fig. 4.1. An improvement on this model was introduced in the form of Riesz's mechanical vocal tract. Its easily recognizable similarity to the human articulation tract gives grounds to the supposition that the flow of air produced by the directed changes in air flow could be transformed into something resembling a speech sound (this historical summary is taken from [5], [71], [224]). The disadvantage of the mechanical constructions was that they had to be operated by a human. This means that the mechanical speech synthesizer could only form one part – if at all – of a human's ability to speak. There is no information available on the quality of this machine. However, we suspect that the hearer had to have a very vivid imagination in order to be able to understand the artificial speech sounds. The pioneers in this field of

Fig. 4.1. Speaking machine of Kempelen from 1791, reconstructed by the Kempelen Farkas Speech Research Laboratory in 2001, Budapest, Hungary (Nikleczy/Olaszy [187])

speech technologies were convinced that the best way of attaining acceptable speech quality was to copy the form and mechanical processes of the human articulatory tract.

This all changed when progress was made in electrical engineering, and electronics also opened up new possibilities for speech synthesis. The main priority was no longer to copy the human vocal tract as accurately as possible. Instead, research concentrated on producing acoustic speech signals that could be easily understood. It was irrelevant whether this could best be achieved by copying the vocal tract, or whether other methods would be more suitable.

Focus was put on an electronic reconstruction of speech sounds, although less attention was paid to the synthetic speech generation than to speech transmission, for example transmission by telephone. In this pioneering group were people like Reis, Bell, Helmholtz, Miller, Koenig, and Stumpf. The first electric synthesizer which could produce not only individual sounds but also coherent sequences of speech was the voder. The principle of a voder is illustrated in Fig. 4.2. This was followed by research and development of various synthesizing techniques (cf. overview given in [244]).

To summarize: The advent of speech synthesis development was mainly interested in the production process of an artificial voice; voice production was the element of quality. Instructions on how to set the individual articulators in order to be able to pronounce what is to be said are given by the user.

The interface comprises information prepared for the speech synthesizer by a person, indicating how electronic parameters are to be set to enable the machine to generate. In other words, what is to be generated is converted by the user into guid-

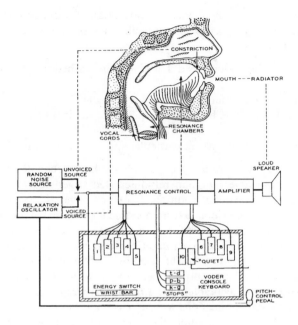

Fig. 4.2. Schematic representation of a voder acc. to Dudley, after Tarnoczy [54]

ing parameters for synthesis. The synthesizer 'merely' generates a voice. To ensure that the terminology is clear, this will be referred to in the following as »voice generation technique«, or referring to the output, as a »synthetic or artificial voice«.

When developing systems the onus was not always placed on analyzing the different aspects of quality of voice generation techniques. In general, comprehension was the main quality feature of each system. Engineers adhered to one particular rule: the higher the level of intelligibility, the better the system quality. As late as 1984 Endres called for greater attention to be paid to measuring speech quality: "It is essential to define the term »speech quality« and to propose methods for measuring it. What we know about »intelligibility« for assessing synthetic speech is not enough." [63]

If we examine this comprehension of synthetic voice and compare it to what has been discussed so far, we come to the conclusion that artificial speech signals indicate far more than just the quality of the voice generation device. They also show whether the information with which the voice generation technology is controlled has quality. In the end a voice generation system can only be as good as the system and as good as the quality of the inputs. This implies that what is understood as

»quality of a voice generating device« always originates from an optimum quality of system input. With respect to speech quality assessment there are certain questions that must be posed:

- What input does the system need that is to be described in terms of quality and quantity?
- Can the quality of the input be described as reliable?
- What relationship is there between the quality of the input and the quality of the output?
- What does the speech quality, in particular speech intelligibility, indicate?

This short précis of the beginnings of speech synthesis technology makes it clear that, at first, the researchers were merely interested in generating sounds which – when they become an object of audition – are processed as speech. In its infancy this interest was relatively isolated from possible applications. Proof of its viability was the main source of motivation, and the first to profit were the scientists who hoped, as Kratzenstein had done before them, to gain insight into either the wider functional connections in the speaking process or other interesting topics of scientific interest. However, the top priority was always to improve the quality of what was already known and available. Using newly garnered knowledge to make advances in products and applications was far less important.

According to the author's opinion, the same can be said for many of today's activities in the field of speech technology. There are still many unsolved or inadequately answered questions for scientists to contemplate. The approaches to determine voice generation technology for example also aim to identify special personalized characteristics in the speech signal flow of individual speakers and to copy them in the synthetic process [40], [218]. There are also unanswered questions related to intonation and prosody, phonetic and linguistic tasks [245], [258]. Scientists are interested in analyzing the basic processes, recognizing, describing and explaining the sequences of events in order to be able to simulate them. Although more attention is being paid to single questions that have yet to be answered, this does not prevent other groups from engineering individual speech technology devices.

In this context there are other influential factors that essentially determine the quality of entities in research and technology: the quality of temporal and financial support and political concerns. In view of the enormous pressure brought to bear by onslaught of competition, both research and technology have come under pressure to produce results. With ever decreasing financial support and time, high quality knowledge should be produced. This means that often those research initiatives are given priority that can come up with quality products, useful both to the individual and society as a whole, in the shortest possible time. Often there is little time for basic research of things like improving the quality of what we already know. As a result enhancing expert knowledge is increasingly coupled with designing products.

The former no longer precedes the latter; both are ever increasingly parallel processes controlled by external demands and, in terms of expended effort, both are kept to a minimum.

Many aspects discussed apply just as much to voice and speech quality assessment as to other engineering products. The quality of today's speech synthesizers or speaking machines is often measured purely in terms of the quality of the speech sounds that are emitted. The only comparison made is with the quality of natural speech. When analyzing system quality, the only question posed is often whether the quality of the system fulfils the quality expectations. Just as there is little interest in what aim the design of a quality entity should fulfil, the general conditions under which the entity of quality should be designed are drawn into the assessment process. However, the following questions which influence both the quality of research and the product design should not be ignored:

- What commitments does the design entity »quality« have to adhere to?
- How much time is available for designing quality?
- What financial support is available?

So far the main topic of discussion has centered on the quality design of knowledge and products. In terms of quality assessment of speech technology components, the listener's expectation (the desired features) plays an important role. This aspect will form the main topic of discussion in the following.

4.2.2 Experience of techniques and expectations of quality

As has already been mentioned, the aim of designing top quality speech technology products will only be a success, if individual quality expectations are also adequately analyzed and applied in technology. If the speech technology products are limited to those that are related to the creation and processing of speech, then we find that the link between individual quality expectations and experience is particularly strong. Experience forms the basis of our daily contact to speech, to different kinds of speakers and their individual ways of speaking. Apart from experience with communication, quality expectations also have something to do with experience of techniques and technological knowledge. This is illustrated in the following example:

When Edison's phonograph for storing acoustic events was introduced, and when by means of this technology it became possible to first preserve acoustic signals and then to call them up as often as was desired, this technique was considered revolutionary. Before that it had not been possible for the listeners to access acoustic signals whenever and wherever they wanted.

Just the possibility of storing acoustic signals was considered to be extremely advanced. The quality expectations of the stored and then accessed speech sounds was of minor importance, because the quality capability of the whole technology had previously been called into question. The decisive requirement for quality was its ability to store and reproduce, a feature of quality that was subjugated to others, such as storage capacity and quality of the speech signal.

Luckily for the system developers, previous doubts on feasibility resulted in a relatively low demand for quality of the acoustic signal. The listener simply did not expect the same quality of sound and speech he was used to in natural situations. For him the decisive factor was that acoustic signals could be stored and later accessed, and not what levels of quality could be expected in this process.

When we talk about the hearer, we do not necessarily mean a layman. Those who have experience of technology are very much aware of the technical possibilities and tend to measure technological progress accordingly. The quality of technology does not only apply to speech quality. One important quality feature is to find an innovative, advanced solution for a particular task, or to improve functionality in an application.

Today's users of sound carriers in our highly technological heterogeneous multimedia society are usually used to CD quality sound. The expected quality is therefore totally different to what was expected at the beginning of the last century. Early solutions are considered inadequate, unless they are seen from a historical perspective. Today totally different standards apply when quality is assessed. The main quality feature of sound carriers for speech recordings expected today is voice and speech quality, mainly because the storage and recording possibilities for this technology have improved far beyond what was originally envisaged.

Today a huge number of technological devices are on offer. The layman can hardly appreciate whether something new is really revolutionary or whether a known process has simply been transferred to a different application. This means that laymen's perceptions of quality in technology often lack depth, and they remain largely unaware of what is innovative, or what is merely a new application for a known method.

This leaves researchers and technologists with a problem: When designing the system technologists who wish to design an optimal, user-friendly system have to analyze and adequately take into consideration what the average potential users' expectations and assessment behavior are likely to be. Only in very limited cases are system designers' knowledge and intuition enough to be able to roughly predict how users will react. This is not because they are unable to put themselves in the position of the user, but because they can rarely prevent their own interests from impinging on the users and cannot put the users' interests first. Sometimes the users' interests are accurately identified, but they then become relative when what is, in principle, possible is compared with the technological aims, for example:

Developers of speech recognition systems know perfectly well that the machine's user ultimately expects the system to have a speech understanding performance comparable to a human listener. This function can usually be adequately described in terms of the features. However, there are certain things that a human can do that a machine cannot. Typical examples of this are the current unfavorable signal-to-noise ratios, processing speech containing strong dialects, processing pathological speech events, eliminating homonyms as well as comprising emotional and connotative meanings.

Background knowledge of the current position and future possibilities of speech recognition research form a baseline against which the suitability and quality of the users' expectations of a speech recognition system must be checked. This quality is assessed with regards to the given possibilities in terms of whether the expectations either at a given or future point in time are technically released. If it is not possible to enhance the system in terms of user expectation, there is a growing tendency to explain away this missing function as excusable, and to gradually move away from the aim of satisfying user expectation.

At the same time experts hope that the user will adapt to what is technically feasible and will concentrate on the central aspects of quality of the existing system (accommodation of the expected quality to the feasible technical quality). The question of what concessions can be made with regard to the loss of quality is often posed because engineers know every inch of their own systems and can therefore anticipate or interpret limited or bad system performance using their in-depth knowledge.

If users do not have access to this expertise, they cannot be expected to be as tolerant regarding the system. In this case the users' expectations are at odds with the anticipated ones. The greater the deviation, the lower the user-oriented effectiveness is.

Even here it becomes clear how important a strict division between research, development and application is. Depending on what state an entity to be qualitatively and quantitatively assessed is in, the assessment will either be global or in-depth, it will be done by experts or laymen, in the laboratory or in situ.

4.3 Elements of quality of speech synthesizers

In previous chapters speech technologies and their associated issues were viewed from the perspective of quality design. Individual systems were mainly handled like »black boxes«. Design objectives were set and illustrated by examples, and the influence of prerequisites and conditions on the feasibility of reaching these goals was discussed.

This chapter will look at a »glass box« approach of speech technology systems, using speech synthesis as an example. One part of this will include a detailed description of system input, architecture and output.

4.3.1 The system input

In general, the term »speech synthesis« is understood as a transformation of information coded other than acoustically into acoustic speech signals. In literature ambiguous terminology has arisen: The term »speech synthesis« is frequently synonymous with »synthesis-by-concept«, »text-to-speech conversion« or »voice generation technique«. Only when synthesis is clearly coupled with a further restrictive human performance (e.g. dialog, transmission or translation systems) is this part of the system clearly defined and named by an internal border (cf. [22], [185], [259], [260]).

In this book the term »synthesis« is synonymous with »speech device«. Synthesis is used as a hypernym when it is not necessary to differentiate between systems of varying performances. If a differentiation has to be made, it is done according to different system inputs:

- translated text-to-speech synthesis
- dialog-to-speech synthesis
- concept-to-speech synthesis
- text-to-speech synthesis
- phoneme-to-speech synthesis
- manually operated synthesis

It is easy to recognize that there are considerable differences between the system complexity of translated text-to-speech to manually operated synthesis. If the system input is formed by a translated text, the number of potential errors is considerably higher than being manually operated by an expert.

The above mentioned classification only refers to the scope of the speech synthesis performance and therefore includes part of the system complexity. The following classification goes in a similar direction [22], [69]. It is divided into:

- announcement systems (acoustic announcement of a text with a limited vocabulary)
- text-to-speech systems (reading aloud a text with an unlimited vocabulary)
- concept-to-speech systems (speech output following a logical and semantic concept)

Automatic announcement devices have a limited store of speech units, e.g. sentences, phrases or words. Today they are mainly digitally stored and acoustically transmitted on demand. The order of the store's entriesis controlled, but not the units themselves. They are optimized by the reference of read aloud known texts[205].

Automatic text-to-speech devices can, in principle, read out aloud every conceivable written text of a target language. It is important to mention here that the processing begins with an analysis of the written text and ends with the acoustic output, in special cases also with electrical output, e.g. coupled to a transmission channel.

A concept-to-speech device processes messages. It has stored special speech patterns for different text domains, e.g. for a weather forecast. When formulating a message, data for speech signal generation are accessed, adapted to the patterns and acoustically generated.

This classification is ordered according to the complexity of the desired system's performance. No matter which classification is chosen, it becomes clear that when assessing quality, it would be totally inadequate to put all the systems into one category of »synthesis« and to assess their quality using the same criteria.

In addition it is also beneficial to split the general functions of synthesis into its terms and definitions and to distinguish between the following:

- speaking machine
- speech synthesizer

When more complex systems are involved in which different speech and language technologies are linked together it is correct to use the term »speaking machine«. These machines consist of various functional entities of which producing artificial speech is just one.

In the following elements of quality of a synthesizer are discussed. One of the most commonly used synthesizers at present is the text-to-speech synthesis. If the aim is to examine the quality of a text-to-speech synthesizer, it cannot be realized without also examining the input text that is to be transformed.

4.4 Text-to-speech synthesis: The text as an element of quality

Written texts form the input for text-to-speech synthesis. The texts are scanned in, processed and finally acoustically generated as speech – in other words the machine reads the text aloud. In order to be able to describe the system performance of a text-to-speech synthesizer, the process of reading aloud also has to be looked at in greater detail.

Reading out loud always involves a text. Texts are sign carriers that lead to associations and meaning at the listener's side. There are different definitions of the term »text«. Here text is used to mean "a functional sequence of spoken or written linguistic elements that constitutes a whole." [91]

Written texts consist of optical signs (unless they are electronically stored). These signs can look very different: In simple written texts they appear as a sequence of letters and punctuation marks, as numbers, arrows, lines or colors which convey a meaning in graphical designs, but also as totally different pictorial elements including abstract and realistic pictures. Stankowski [239] differentiates between the following basic forms employed in visualization:

- reality and its images
- abstract illustration of images of reality
- symbolic representation of reality that cannot be mirrored
- pictorial representation of functions
- representations of systems

- representations of codes
- technical representations
- artistic representations

Having identified the representations of codes, the process a person reads a text is as follows: "In its most general form, reading can be defined as a visual sampling of a two-dimensional representation that is not − or not only − perceived as the surface, but as a sign carrier." [91]

The sign carrier text conveys a sense, when being processed it leads to the association of meaning. The characters of the text stand for something else. In contrast to reading a text »reading« a photo is something different again: A photo − the surface of which also has features similar to a text and can be optically scanned by the observer − is not normally perceived primarily as an arbitrary sign carrier [152]. It can be accessed by every observer without further explanation. However, whoever wishes to understand an arbitrary sign carrier »text« must be able to work out the contents of the perceived surface. To be able to do this the observer has to know the sign conventions on which the text is based. If the conventions of the relationship between the sign carrier and the contents of the signs are not known, comprehensible texts (i.e. sign carriers) cannot be produced, nor can texts be processed according to the conventions they were produced. Put in another way: If you do not know the reception conventions you cannot understand a written text.

But what do these reception conventions look like? In developing text-to-speech synthesis the aim is to copy the human ability to read out aloud and to pass on information acoustically. If one looks at the present performance capabilities of text-to-speech synthesis, one has to admit that this is an ambitious goal. For many different reasons today's text-to-speech synthesis still treats input letters as independent objects of their own. But a letter is, as mentioned above, a sign carrier or a component of a sign carrier. If a machine reads something out loud, it is often reduced to a transformation process in which the sign carrier »written text« is transformed into the sign carrier »text read aloud«. In this transformation process the contents and functions of the signs are largely ignored. The text module is therefore not treated as the sign carrier. What is changed is the surface of the sign carrier without semiosis being involved, and an optical sign carrier becomes an acoustic one.

Even if persons read the text out loud, they change a written sign carrier into an acoustic one. But in contrast to speech synthesis this happens indirectly: While persons visually scan the optical presentation »text« they first understand the meaning of the sign carrier and then produce the acoustic sign carrier, applying this to the production conventions they know to allow the speech to be read aloud. Reading aloud goes hand in hand with comprehension and association of meaning. Main text production and perception processes involved are schematized in Fig. 4.3. for the human reader and in Fig. 4.4. for speech synthesis.

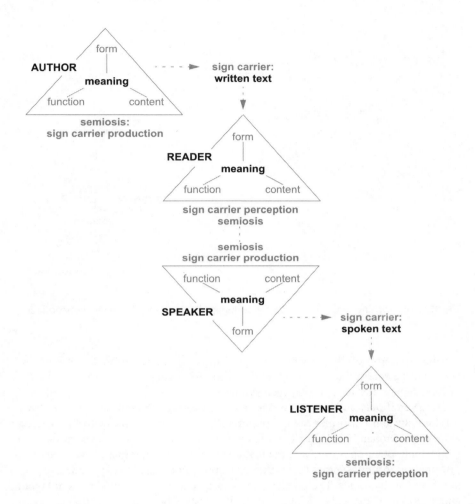

Fig. 4.3. Text writing, reading aloud, and listening: processes of semiosis

By analogy with the above quotation on the natural reading process, the way a text-to-speech synthesis reads aloud can be described as follows: It is an electronic scanning of a two-dimensional optical presentation which is treated as a surface and not as a sign carrier.

In speech technology the term »text« is usually not used in a highly specific way. If speech engineers talk of »text« in connection with text-to-speech synthesis they mostly mean a written pattern of speech in which speech units (e.g. letters, syllables,

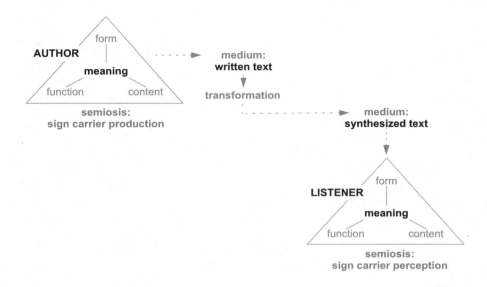

Fig. 4.4. Text writing, synthesizing, and listening: processes of semiosis and transformation

words) are purposefully presented in coherent order [109], [160]. These kinds of texts will be referred to as »ideal texts« in the following (»ideal« in terms of the present capabilities of text-to-speech synthesizers).

But even if only an ideal text is put into a text-to-speech synthesis system, many different system performances are required that are comparable with the reading ability of a human being: Practiced readers know the rules of the system through which their attention is drawn to the features that, in their opinion, convey the most information. Amongst other things this is an ability connected to experience, expectation and probabilities. The practiced readers have learnt the ground rules of layout and reception of an ideal text. These principles ensure that communication is encoded efficiently and meaningfully. Knowledge of these ground rules directs perception. Reading is not simply making a row of individual elements. In the personal reading process:

> "… new, larger units are created that constantly mesh to form new, effective units. [...] The elements of an utterance organize themselves and are functionally effective as new units – this is psychological dynamics [...]." [109]

Ideal texts refer to situations from the life and experiences of the recipients. This common basis helps to prevent obstacles to communication arising in the reading process, such as double meanings and vagueness.

"[....] the readers' attitude towards reception is that they expect it to be possible to apply a consistent meaning to the text." [91].

Thus readers are always, to a certain extent, active recipients who activate information via semantics, syntax and pragmatics in order to be able to attribute a meaning to the message they are reading aloud. This constructive role comes to the fore when literary texts are processed.

"A literary text has elements that are irrelevant or disruptive in everyday speech, but are an essential element of the genre: multiple meanings, different levels of context, and interpretative possibilities, emphasis on the linguistic form. The reader of a literary text expects neither economy nor transparency [...]." [91]

There is a great deal of flexibility involved in reading a literary text. Just as it is impossible to clearly predict individual text understanding and interpretation, it is equally impossible to place receptive behavior in categories ranging from correct to incorrect. Despite this there are certain ground rules that cannot be broken.

Nowadays nobody would seriously entertain the idea of allowing different literary texts to be read aloud by a machine without having first edited them. There are still too many potential pitfalls to get lost in issues on how to model reading idiosyncrasies, emotions and moods that are essential parts reading literary texts aloud. Despite the fact that today there is multiple research going on to acoustically model emotions in speech this is still related to voice generation technique mainly, and only in rare cases to text-to-speech synthesis.

Many speech engineers remove semiotic considerations from their models of reading-aloud processes. They consider the text to be only the medium, a means to an end, and therefore not an independent sign carrier. They first want to model the general characteristics of the reading process, in which the medium (in the case of speech synthesis the artificially produced speech) does not act as a distraction [91]. Their aim is to relegate the form of the sign carrier to a background role, provided that these elements are clearly identifiable and oriented towards pragmatic issues. Or, in other words, their focus is on form characteristics of the generated signal.

4.5 Synthetic speech as a causal sign

In text-to-speech synthesis the form of the sign carrier is an important criterion for speech quality, not only with respect to the system input (as discussed above) but also to the system output. The requirement that a text should be unobtrusive when it is put into the system can be easily met by only choosing ideal texts for synthesis. In contrast it is much more difficult when applied to the system output, i.e. when the synthesized speech signal itself is analyzed from the point of view of it being a sign carrier as well. As the acoustic text is a medium, a sign carrier itself, it should, of course, be unobtrusive and not form a distraction either. When listening to speech the auditory perceived signal should hint at the form–content–function relationship, not at the form itself. Certain forms hint at certain contents and functions, and in turn, certain contents and functions require specific forms. The materialism of the sign carrier »synthetic speech« can only be located, or remain, in the background when it does not catch listener's attention with regard to a violation of this form–content–function relationship.

For many text-to-speech syntheses there is a marked discrepancy between form criteria of this sign carrier and the contents and functions correlated with the signs, in other words, between the acoustic-synthetic form and the contents that should convey it and the functions that should perform it. This discrepancy stems from the fact that the present capabilities of many text-to-speech syntheses are limited to treating the text as a lifeless, sober medium of information. Not only is the relationship between form, contents and function ignored, but the surface design itself has some features that take some getting used to. As a result the quality of a synthetic text turns out to be something like this:

> "However, almost all synthetic speech, although understandable, sounds quite unnatural and highly over articulated. Since the intelligibility of most text-to-speech systems has improved considerably during the last ten years, attention should now be directed more and more to questions of naturalness and acceptability. But up to now it is far from clear yet, which aspects of the speech signal are responsible for the shortcomings in this field." [151]

To a certain extent, this is valid still today. For automatically generated acoustic speech signals there often does not seem to be a communicative relation. The communicative function is solely reduced to purely receiving information [158]. Following Cherry it is possible to say that synthesizers do not provide communicative signs, but rows of causal signs. Cherry [39] distinguishes between »causal signs« and »communicative signs« as follows: Causal signs are nothing more than a predetermined order of physical conditions that require that neither the sender nor the receiver has to behave cooperatively. Information is passed on without either party influencing the process. The sender and receiver are not independent of each other. In a strict sense they belong to the channel and are therefore carriers of information.

Communicative signs refer to an active sender who wishes to communicate something, a channel through which the message will be sent, and a receiver who will be reached by the message and who will process it.

> "[...] the distinction between direct causation [...] and communication [...] is that the first is a simple, inevitable, cause-effect relation, whilst the second is only a probabilistic cause-effect relation." [39]

Therefore the text read aloud by a machine is, similar to a written text, only a medium, a means to an end. The transferal of information takes place without either the sender's or receiver's information being influenced. The sender could be a measuring instrument that provides the input data for the concept-to-speech synthesis (e.g.

measured temperature for an automatic weather forecast synthesis system), or an author who writes down a text which might be then processed by a text-to-speech synthesis system. When these different input types are processed by a speech synthesis system, its output speech signals are causal signs, because the predetermined system inputs do not require a sender nor a receiver to behave cooperatively. The information reaches the perceiver in the form of acoustic signals, after it has been transformed by the machine. What constitutes the written text for the reader is synthesized speech for the listener. In fact it is only the surface design of the text that has been changed. Both the optical and acoustic texts constitute a mass of signs that have the function of passing on information.

As synthetic speech signals are causal, whereas natural speech sounds are communicative signs, synthetic speech is not to be processed like natural speech. Persons who listen to speech generally expect natural speech to be the object of perception. In other words, they expect communicative signs and a speaker who wants to communicate something – a speaker who actively communicates with the listener. With artificially produced speech, the active communication process is interrupted in two ways: The speaker and listener have only very limited, indirect, anonymous 2-step contact with each other via (1) a written text and via (2) synthetic speech.

To be able to describe listening behavior, or even the attitude to synthetic speech, it helps to imagine how persons behave phonetically and linguistically when they speak to a machine, looking specifically at those speakers and listeners who have relatively little experience with communication technologies. The aim is to draw attention to spontaneous speaking and listening behavior with speech technology involved to which users have not yet adapted. When perceiving synthetic speech, there occur behavioral reactions that are comparable to the experience of phoning someone and unexpectedly being connected to an answering machine. The speakers know that the machine cannot understand them and that it is merely storing the message. They also know that the intended recipient will listen to the message later, but they still perhaps have difficulties in expressing themselves as spontaneously as they would if the recipient of the message were on the receive side. This knowledge affects their speaking behavior: They speak more slowly, stylized, and communicatively reduced.

This indicates that if we know that the source and sink of the message are separated spatially and temporally, it leads to an observable change in speaking behavior. This way of speaking is often described as unusual or inadequate, when seen in a communicative context. Speakers attempt to produce causal signs, but the way they express these signs (which will ultimately be perceived as a communicative sign) differs from the normal communicative sign. Just as speakers have problems producing causal signs, listeners also have problems processing them. The listener cannot help but notice the stylized way the caller speaks; form characteristics come to the fore.

With synthetic speech, the source of the message and the listeners are also spatially and temporarily separated. The listeners know that the information is emanating from a machine and they notice the ways in which the synthesized speech sounds differ from natural ones, specifically when the reference is a direct speaker-listener constellation. Despite this they experience difficulties in reception, for example because the way in which the information is being presented to them (form–content–function relationship) is impaired.

When reading aloud, the speaker (be it a person or a machine) is part of the channel, because the »real speaker« (here in the sense of the source of the message) is the author of the text. The listeners are accustomed to the reader being cooperative and communicatively active. The reader does not simply convert the written text into an acoustic form, but attempts to prevent any discrepancy occurring between the way the message and its contents are orally transmitted (refers to the function of sign carriers). The speech signals are perceived, understood and interpreted accordingly. Speech, both in speech production and reception, is always communicative behavior.

Listeners know the conventions that go along with natural speech processing. They know that natural speech is individual, intentional, and conforms to certain rules, but is yet infinitely variable. Therefore the listeners expect both variation and uniformity. They only expect as much variability as the language system will allow, and uniformity must also confine itself to those aspects which the system is based upon [67]. Jakobson/Waugh describe it as follows:

> "The two essential focuses of linguistic investigation, one upon invariance and the other upon variation, are sterile separately from each other, and any one-sided exaggeration – one might even say monopolization – of one of the two facets with a disregard for the opposite one distorts the very nature of language. Any system is by definition mutable, hence the notion of an individual or collective language system, without variation, proves to be a contradiction in terms. The notion of context which furthers variation becomes even wider and encompasses not only sequential and concurrent neighborhoods in the sound flow but also the diversity of speech styles." [124]

Variability and uniformity are therefore characteristics of the system that need not be employed arbitrarily.

> "Different varieties are appropriately used in different contexts in a speech community. In this way, a set of linguistic expectations is created among interacting speakers, and a set of reciprocal rights and duties emerges. But a given context demands the use of a certain variety, the use of an unexpected variety will be interpreted as a meaningful choice, and an indication that the speaker is trying to alter the context. Rather than attributing the unexpected choice to a lapse in the speaker's attention [...] the interlocutor will attribute

it to the speaker's intention, and will try to derive a new meaning in order to interpret the new context adequately." [4]

When analyzing texts read aloud by a machine, it is directly noticeable that the rules on variability and uniformity are not always rigorously adhered to. There occur variations in the speech style that result in an inadequate interpretation of the utterance at the receive side. This often leads to communicative interference. Obviously it is not easy for the listeners to disregard properties of the synthesized speech sound so that they can ignore changes to the speech style and cognitively treat the surface as a causal and not as a communicative sign.

Present-day synthesizers are unable to generate a way of speaking that constitutes the most suitable variable for a certain communicative purpose or particular contents [38]. Synthesizers are incapable of modeling all the ways of reading to include all the linguistic and phonetic speech style variables in relation to the contents [64], [65]. They cannot generate communicative signs.

"Although some TTS systems perform quite well at the task of reading aloud information in a »neutral« speaking style, the production of completely natural-sounding speech and the extraction of higher-level (semantic, pragmatic) information from orthographic text, and its appropriate realization in terms of acoustic parameters, remain major challenges for the future." [201]

Even today this deficit has not been solved. From a theoretical point of view one possibility is to reduce all the conceivable, individual speech style variables to suitable prototypes and to base speech synthesis procedures on them. In fact, to a certain extent this is also the ground for signal manipulated time-domain synthesis; however, time-domain synthesis has developed into another direction [154]. The requirement of a prototype-driven speech synthesis is that it is skillfully introduced into social contexts because listeners need to learn appropriate reception conventions. Such a synthesis would create a new acoustic text type which has its own rules, but − and this is highly important − does not interfere with the production and reception rules of natural speech.

In principle this already happened when written speech was introduced with the written text as the new carrier of information. But, in contrast to written speech, synthetic speech is a sign carrier that bears a strong resemblance to an existing well-established sign carrier, namely natural speech. In terms of semiotics this means that today we are dealing with synthetic speech as a sign carrier that does not function independently, but is an indexical sign that refers to natural speech, and natural speech is in itself an arbitrary sign carrier.

One long term minimum aim is to record the most important speech style variables and the rules of speech styles by analyzing natural speech, naming the necessary conditions and using them for technical copies. To realize this it is necessary to first destroy the dependency of phonetics and prosodic units in speech signals from the syntactic, semantic and pragmatic text units. They must then be classified and their effects assessed.

Numerous analyses, descriptive and explanative studies have already addressed this topic. So far they have not split up into reliable elementary events that can be effortlessly generated in synthesis systems [175]. There is still no possibility of recognizing text types by machine that require special speech styles in order to be able to control the synthesis behavior using this information. Text types include

"[…] argumentations, letters, discussions, […], instructions for use, descriptions of appliances, interviews, sermons, adverts, recipes, radio news broadcasts, job advertisements, statements, telegrams, telephone conversations, lectures, weather reports, newspaper reports." [160]

4.6 Functional internal system entities as elements of quality

In information and communication engineering, phonetics, linguistics, and communication science there are many research projects dealing directly or indirectly with specific aspects of speech synthesis. We have already mentioned voice generation techniques. In this section will be presented other internal functional entities that are necessary if written text is to be transformed into speech.

Of course speech quality assessment does not only cover assessing synthesized speech sounds, but also examining the range of performance of individual functional entities of syntheses, e.g. graph(eme)-to-phoneme conversion, stress assignment, text and sentence analyses. To achieve this, the expert on quality must, at least, have a rudimentary knowledge of how the various processes work. As not all the activities are based on one system, the individual components of different systems are not necessarily interchangeable or compatible. If that were the case, it would be relatively simple to directly compare the advantages and disadvantages of the components to be assessed in the conversion process. The quality of the individual components is often only apparent after several functional units have been executed. This makes it much more difficult to clearly detect poor or non-performance. What synthesis techniques is concerned, it is useful to differentiate between the elements of synthesis and the types of synthesis:

- elements of synthesis: phones, diphones, demi-syllables, syllables, words, sequences of words, phrases, sentences, paragraphs, texts
- types of synthesis: either directly based on the signal form (so-called time-domain e.g. PSOLA) or the signal form is parametrically labelled. Synthesis occurs via the parametric representation (so-called parametric synthesis, e.g. formant or LPC-synthesis)

An infinite number of elements of synthesis can be employed in both of these processes. The choice of elements is made according to the criteria related to the task and the technology. For example:

- limited versus unlimited vocabulary
- much versus little storage capacity
- large versus small corpora

Discussing this exemplary case makes it clear that when assessing the quality of synthesizers it is also important to distinguish between

- the quality assessment of the system output (synthesized speech)
- the quality assessment of the performances of certain system sub-entities.

4.7 Speech synthesis and associated goals in research

Generally the aim of researchers is to extend knowledge and thereby facilitate progress. The key to achieving this is flexibility and a supply of good ideas. But often the prerequisite cannot be realized because too much emphasis is placed on what has been achieved and what is presently available. Often there is not enough flexibility and courage when it comes to going right back to earlier steps or even to the beginning to deal with the basic issues once again, but from a different perspective.

This is also the case with speech technologies: If we systematically examine the pros and cons of all the possible system structures, the bases of our knowledge and process control, it will become clear how we should construct the best possible system to deal with today's requirements. However, already at 1984 Fant regrets that activities concerned with speech synthesis are all too often concerned with implementing a functional system and that enhancing knowledge is less important [67]. This is often the case still today. Pointedly going back to the basics and dealing once again with specific issues can increase the necessity to find new solutions. This can have far-reaching consequences for system performance, e.g. for the quality of synthetic speech: Basic changes can, at first, lead to a considerable loss of quality of the system output, which, however, in the long term will facilitate progress. What is a loss of quality for the system output, may signify an enhancement of scientific knowledge and may therefore be an advantage for the future product design.

There has also been a call to thoroughly revise the methods and methodologies of speech technologies. This is justified when you consider the relatively limited performance possibilities of today's synthesizers in view of language-in-use:

"[…] speech can only be really analyzed as a concrete »language-in-use« in the context of real actions. Every attempt to comprehend speech as a system of means free of applications that can be implemented as and when they are needed is doomed to failure. […] if language-in-use is to be examined, then not only the speech processes should be looked at […]." [169]

A synthesis that takes into account the communicative processes in typical fields of action requires the inclusion of further sources of information and perhaps even of system structures other than the ones that have so far been investigated. The process of reading aloud is a reading and speaking action. It is also a mix of speaking activities that have their own subordinate aim. This is, in turn, subordinate to the motive of the action [159].

Reading aloud is therefore a speech act-in-use. To achieve a suitable, technical imitation it is first necessary to redefine the object of analysis in the context of the practical application situations. It can be assumed that the object of analysis, reading aloud, is a typically active functional unit in which can be found sufficient specific speech act patterns that can flow into the synthesis process [169].

If the goals of research in speech synthesis were to be readjusted, the initial result would definitely be a loss of speech quality. The reason for this is the extreme shift in focus, which is especially helpful when it is a case of discovering new perspectives. A shift in emphasis can enhance knowledge and therefore decidedly improve the functionality of synthesizers in the long run. However, the chance to carry out a comprehensive assessment of scientific progress often goes to waste because individual performances of sub-entities in the whole system are not adequately linked into the corresponding quality picture. This could be the above-mentioned scientific question that refers to how to technically model communicative processes in a typical, communicatively active scenario. It is without doubt unreasonable to take such a synthesis out of its research context and determine its quality by means of an unrelated or poorly connected assessment of synthesized speech. Consequently we reject Edward's view that "the term quality applied to synthetic speech refers to its closeness to natural speech" [57].

If the quality assessment sets out to analyze the entity to be assessed according to its aims, this is even more difficult than usual for research: An adequate assessment often fails to materialize because the quality of the documentation on the idea, specification, design and execution are either inadequate or non-existent. If the intended goal and motivation for setting this aim are not clearly described, it is hardly possible to compare and assess what has been achieved with the original aim. As the above has proved, this approach cannot be omitted if the assessment is to have a solid basis for comparison.

So far there have been few attempts to systematically analyze speech behavior in different active situations. Trautmann/Langhoff, for example, tried to construct speech behavior in a working environment under different conditions [246]. The experiments done aimed to systematically examine how direct speech communication is impaired in order to obtain an instrument that could be used to assess and design workplaces with regard to the demands of communication. From the perspective of health and safety at work it deals with speech quality in different working environments, which can be used to influence productivity. Within this framework the following questions were the focal point:

"How large are the acoustic parameters (noisy environment, acoustics in the room) allowed to be so that the speaker can still communicate the information necessary for carrying out the activity at a suitable vocal level and still be understood? Which parameters, irrelevant of the work activity, favor or aggravate comprehension? What quality of speech communication (conversations with normal/limited vocabulary, simple orders, verbal warnings, instructions etc.) is necessary for each activity and by which parameters related to the activity is it influenced?" [246]

Regrettably the influential factors identified are subsumed under the term »communication qualities«. However, this term covers far more than what Trautmann/ Langhoff intended. Moreover, the features of communication quality and its systematization picked out by the authors leave both room for improvement and a number of questions unanswered.

Despite these weaknesses the basic idea is very interesting for the quality design of speech technologies in daily use. The idea that man has access to a number of features of speech-in-use that are motivated by language-in-use, opens up new possibilities for quality design. This is a case for linguists and communication scientists.

At present there is no suitable classification of speech-communicative processes of typical fields of action that is beneficial for designing practical synthesis. As in situ experiments and analyses are time consuming and expensive, and the object being examined varies greatly, quality design is often oriented towards very general conditions. The aim is to define enough features of the application for which the synthesis system has been designed, and to design a scenario for the development that has the same features as the application. It is also hoped that the optimized system behavior can be transferred to other applications without a great loss in quality – the designed elements of quality are therefore distinctive for the system quality in the whole application. This is the only way of ensuring that the designed quality has at least some similarity to the designed quality of the system in use.

The long term aim of quality design is to compile a descriptive system of elements and features of quality that can be used to develop a well-based classification scheme of conditions and requirements of quality in activities. This scheme is comparable to a topographic network through which variables and structures of a variety of active applications can be revealed, which are then reintegrated once more into designing and assessing the quality of speech-in-use.

4.8 Summary

With respect to the group of listeners/test persons/customers the following aspects were discussed:

- What is the listeners' general attitude to technology and especially to speech technology?

- How much knowledge do listeners have of speech technology?
- What experience of technology do they have?
- How do listeners assess technology's potential beforehand?
- How do they assess improvements in performance?
- How do they assess a system's degree of innovation?
- What do they consider to be the most useful features of new technology?

With respect to the group of researchers/system developers/suppliers:

- What are the aims of researchers, system developers and/or suppliers?
- Which difficulties immanent to the task do they have to face?
- Are there tasks to which there are principally no solutions which have an effect on the intended product quality?
- Do the researchers/developers limit their research/development/optimization to certain sub-functions of the system?
- Are partial short term solutions necessary? If so, which ones?
- How do such unavoidable short term solutions manifest themselves in the system's behavior?
- What loss of quality has to be accepted because of this?
- Are any of the expected concessions to quality adequately described?
- Is the system's functionality sufficiently documented?

With respect to the system classification:

- Which system is involved?
    - speaking machine
    - speech machine
    - translated text-to-speech synthesis
    - dialog-to-speech synthesis
    - concept-to-speech synthesis
    - text-to-speech synthesis
    - phoneme-to-speech synthesis
    - manually driven synthesis

With respect to the system in- and output:

- What features are inherent to the input (e.g. complexity of the text)?
- Which speech style does the input require?
- Which features does the system output have (synthesized speech)?

With respect to speech communication processes in applications:

- What is said?
- How is it said?
- To whom is it said?
- How is the message passed?
- Where and by whom is the message received?

- What are the listeners like?
- Which aspects of quality have top priority?

Describing the goal of speech quality assessment:

- quality assessment of a research system
- quality assessment of an application system
- quality assessment of sub-entities of a system
- quality assessment of the whole system
- quality assessment of functions of a system
- quality assessment of the effect of a system

5 From Speech Perception to Assessment of Quality

Speech quality is, in a mathematical sense, a function with several variables. It is related to perception, motivation and attentiveness, subjectivity, perspectives, selection, construction and memory, knowledge, experience, expectations, the brain and the nervous system (cf. [105], [208]).

To remove the anonymity surrounding the assessment process, important aspects illustrated by models will first be discussed by posing epistemological and theoretical questions generally and measurement-based questions specifically. The following model for auditory perception is taken as the basis (cf. [93]):

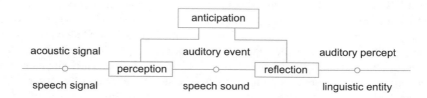

Fig. 5.1. Model of auditory speech perception acc. to Blauert/Jekosch [20]

5.1 Perception and cognition

Before carrying out voice and speech quality assessments, one first has to consider the prerequisites and conditions of perception and cognition. Irrespective of the type of object to be perceived, there are various attempts to explain the processes of perception and cognition that relate to the basic question of how we perceive the world and how we recognize and understand it. The scientific field that deals with comprehension, cognition and knowledge is epistemology. Epistemological theory provides particularly fruitful impulses for this work. Anyone who answers the question of: "How does a person perceive the world?" will first refer to the sensory organs. This leads to the following epistemological dilemma:

"We think we know what »sight«, »hearing« and »touch« are because perception always involves coloured objects or things that emit sounds. However, if we choose to analyze »sight«, »hearing« and »touch«, we then transfer these objects into our consciousness. We commit what the psychologists term the »experience error« by attributing what we know about things to our own consciousness of the objects. From what is perceived we create perception." [174]

We know something about the world because we live and process signals that are offered to our sensory organs. We perceive these signals by means of our sensory organs. The term »by means of« is not intended to be understood instrumentally, namely that the sensory organs act as a purely anatomical guide and converter in that they register information being submitted, pass it on and store it as perceived information. »By means of« is meant here in the sense of »to be on«.

Perception is an active, individual process. It generally presupposes that the subject of perception receives signals and that it is very much capable, from a physiological and neurological point of view, of perceiving such signals. But as a matter of fact, it covers much more.

Of course perception is basically made possible by the signals on offer and the prevailing physiological, peripheral conditions. Perception actually occurs in a certain environment, in the context of experiences, education and knowledge, and is determined, for example, by emotional and cultural factors. Knowledge and experience draw the attention of the subject of perception, even before perception takes place. Knowledge and experience control the selection of signals arriving and dominate processing and interpreting the signals. The sensory organs constantly receive signals. Only those that the perceiver can at that moment in time grade as a potential source of information will be concentrated on and processed. In this context expectation and anticipation play an important role:

"We think of perception as an active process of using information to suggest and test hypotheses. [...] By building and testing hypotheses, action is directed not only to what is sensed but to what is likely to happen, and it is this that matters. The brain is in large part of a probability computer, and our actions are based on the best bet in a given situation. The human brain makes efficient use of its rather limited sensory information [...]. If the brain were unable to fill in the gaps and bet on a meager evidence, activity as a whole would come to a halt in the absence of sensory inputs. In fact we may slow down and act with caution in the dark, or in unfamiliar surroundings, but life goes on and we are not powerless to act. Of course we are more likely to make mistakes (and to suffer hallucinations or illusions) but this is a small price to pay for gaining freedom from immediate stimuli for determining behavior, as in the insects which are helpless in unfamiliar surroundings. A frog will starve to death surrounded by dead flies." [88]

Not every signal leads to perception, and even if perception occurs, the result of the perception process is not absolute (cf. also [173]). Perception is relative, as it is subject to circumstances.

"One »component« of perception is always in the realm of another, always a piece in a »puzzle«. Nobody would be capable of making an absolutely homogenous surface on which there was nothing to perceive the object of perception." [174]

The following concept can be deduced from this discussion: Perception is a process. In the auditory field, it is the auditory event that takes place. The perceptual event is an individual fact, therefore phenomenal. It can only be accessed by means of reflection, i.e. in an intellectual form. As soon as the perceptual event becomes a topic for thought, the term becomes known as what has been (reflectively) perceived. Then what has been perceived is a fact of description, and is therefore phenomenological.

Several individuals may describe their perceptual events in a similar way, despite their subjectivity. This fact is of great importance to the process of speech quality assessment. Speech quality assessment aims to analyze the typical aspects of what we hear. Supposing that an inter-subjectively comparable description of objects of perception can be called the typical aspects of what has been heard, we also have to understand the contribution, kind and expression of intra-subjectivity, so that we can comprehend the processes of voice and speech quality assessment.

5.2 Perception between individuals: Form and adaptation

All individuals are able to recognize objects that they might have seen before under very different circumstances, e.g. light, distance, position. They are able to identify people who have obviously aged after not having seen them for many years, can recognize the identity of different forms that have roughly the same size, and can predict the way well-known but transposed melodies will continue.

These are clear signs that perception occurs in a network of complex signal processes involving experience, cognition, and knowledge. Perception may be unique and governed by the context, but it is in no way uncontrolled, chaotic or even haphazard.

Our sensory organs are constantly flooded with signals. In order for the individuals to have any chance of orientating themselves in this chaos, the range of signals on offer is put into some sort of structure by the perceiving subjects. As already indicated, the perceiving subjects seem to concentrate most on those signals that contain the greatest amount of information that is of particular interest to them. Perception occurs therefore against the background of what has already been experienced, learnt and recognized. Thus perception is clearly dependent on representations [148].

"We know that there must be [...] representations of past sensory experiences
[...], emotional experiences, of experiences of movements [...] and even of
episodes of interpersonal experiences [...]. But we really know very little
about how all these kinds of experiences [...] are actually represented in the
long term memory." [265]

What do we know how these representations look like? We know that the perceiver
does not only comprehend the appearance of individual objects and events but also
(or especially) the relationships between objects and events. If perceptual events
take place, the information on perceived objects will be relentlessly collected and
updated. This knowledge can be recalled at a later date.

There have been many theories, investigations and attempts to explain the question as to how things that are perceived are represented in the individual's mind. The
main means of collecting supporting data is self-observation or introspection.

Within the framework of quality assessment it is of interest to analyze more
closely the perception process of individuals. How does the perceiver structure the
signals on offer? The strategic method used to determine perception is specifically
geared towards quickly and economically discovering invariance in the chaos of the
signal varieties against a background of a set of representations [147]. This intricate
procedure enables objects of perception to be conceptualized. In the context of epis-
temogenesis Vollmer [253], for example, mentions three aspects of invariance: the
constancy of objects (recognition of objects despite different distances and perspectives, including constancy of size and shape), the constancy of direction and the constancy of color. In fact aspects of invariance are closely related to the concept of
»Gestalt« (or »whole«). Gestalten (or wholes) have first been thoroughly investigated, e.g., by Christian von Ehrenfels (1859-1932) [66], Max Wertheimer (1880-
1943) [165], Wolfgang Köhler (1887-1967) [106], and Kurt Koffka (1886-1941)
[146]. Generally spoken, Gestalt theorists take up the question of why we perceive
and organize objects the way we do.

The most important ones for creating optical figures are the laws of proximity,
equality, unity, of a constant curve, common fate and experience [142]. Von Ehrenfels established criteria for melodic patterns (cf. [233]). He draws on the fact that a
melody is more than the sum of the individual notes (so-called »Übersummenhaft-
igkeit«) and mentions transposition as a second, important criterion which is typical
of melodies.

The common conviction held by all the Gestalt theorists is that the range of various signals does not gradually lead to individual perceptual events themselves, but
only enables perception to be autonomously organized. However, this organization
obviously has its own set of rules in which the various performances of the perceptual system are integrated and abstracted. In genetic epistemological theory auton-
omous systems like these are called schemas [202], [203].

In connection with speech quality assessment it is particularly important to know how perception occurs when an unusual signal is to be processed. Perceivers expect both coherence and transparency and want to use their knowledge of relationships. But what happens when something unpredictable is to be perceived, if the signal on offer breaks conventions or does not fulfil expectations?

If the perceiver is confronted with a signal that he/she has never encountered before, a disequilibrium may occur between what is perceived and what has been learnt or what is expected. This imbalance has to be corrected. According to Piaget two strategies are possible [6], [202], [203]:

- **Assimilation**: The new offer is to be aligned with the existing schema. It is put 'into another code' and integrated into the existing system.

- **Accommodation**: The existing schema is changed to suit the conditions of the arriving signal. The schema remains flexible and adaptable until a possibility for integrating what is to be perceived and what has already been perceived has been found. New schemas may develop.

In the course of many investigations Piaget and others have been able to prove that usually both processes control the establishment of an equilibrium. As soon as the perception process is only assimilating or only accommodating, a pathological case is given [79]. These two strategies for adapting will be discussed in more detail later.

Before it must be emphasized that the perceptual event is the result of a comprehensive context in special configurations of situations. However, observing an individual's behavior in different configurations of signals shows that the perception process is strategic in that it concentrates on invariable phenomena, and special strategies are used to conceptualize objects. Amongst other things, this supposition is supported by the rules observed in epistemology. In epistemology development occurs in continual processes of discrimination, construction and integration succeeding phases, stages and steps that are irreversible and to a certain degree universal [35], [202], [203], [204].

However, keeping in mind that different individuals have comparable perceptual events, one question still remains unanswered: Are such strategies, used to form invariance in perception processes, highly individualized or do different individuals use the same strategies? In other words: Are each individual's personal strategies of perception rather accidentally formed, but then later given the impression of being uniform? Or can the strategies also be found when a comparison is made between individuals, thus at least demonstrating some degree of universality? There is ample proof that concentrating on comparable invariance phenomena in the perception process is an inter-individual fact [133].

5.3 Inter-individual perception

To be able to learn more about inter-individuality in the speech perception process, it is first necessary to recall once more the connection between the perceptual event and the percept. It has already been said that the object to be perceived in its – theoretically conceivable – original form is not immediately comprehensible, because every attempt to analyze the perceptual event presupposes reflection. By reflection, the real experience of perception is interpreted and thus given intellectual properties. Perception is thus forced to become something that is perceived, in other words, it becomes a factual description. This is also the case with introspection.

If one wishes to investigate strategies for forming invariance in speech perception processes of different individuals, a number of methods could be used. One possibility, which is mainly used in the domain of speech quality assessment, is to give the perceivers the task of describing each object they have perceived. Then each individual's description is analyzed and interpreted, taking into account the original questions posed.

How can a speech event that has been perceived be described? Which processes does the description of auditory speech events go through? For example, if one asks different people to concentrate on the same speech signals and to describe what they perceive, there are often factors common to all the descriptions. Can we conclude from this that different people perceive the same things? Do not they perhaps only give the same descriptions?

What is perceived by another person only exists in the form of a description. The form in which something that has been perceived is communicated is the result of an intellectualization. In order to be able to talk of isomorphs or of the universality of what is perceived, we must first test whether the individual descriptions are also identical to the corresponding – individually reflected – perceived objects. As outlined above, obtaining total and irrevocable proof of this is impossible. If it were possible, the term »objectivity« could be defined, as it is often used:

> "We can speak of objectivity when statements are made concerning a reality
> that is independent of an observer, whereby this reality comprises objects in
> a pure state and can be understood in the same way by all the perceivers. Cri-
> teria of objectivity are: a person's independent statement, the language of
> communication, a reference system, methods and conventions." [253]

Seen from this perspective, such an understanding of objectivity is hypothetical. For the purposes of this book objectivity is understood as the following:

> »Objectivity« is the asymptote of the invariance with respect to the totality
> of what is individually perceived. Objectivity is not absolute, it is not a state,
> but it reveals within a process. Objectivity is the extent of inter-individual
> agreement.

This shows that objectivity and structural isomorphs play a central role in the assessment of voice and speech quality. It is not appropriate to discuss here which causes of structural isomorphs can be cited. In the continuum between a science that aims at objective insights that postulate the possibility of making observations independent of the observer, and a science that understands perception and cognition as a product of a physiological-psychological process, every conceivable explanation is possible. At present it is not possible to give a clear explanation free of contradictions. Every simplified thought of a model seems to limit the arbitrariness of interpretations, but it is impossible to measure how well the model corresponds to the real procedures.

As the question of causes is always to a certain extent speculative, and not every answer that is given can be unconditionally proven, it suffices to say here that there are observable isomorphs in the description of what is perceived, therefore also in the speech that is perceived. It is also important to note that these isomorphs cannot be compared to reaction norms. Any deviation from the expected reaction should not be disqualified on the grounds that it is atypical. It is, in fact, this individual behavior that sets normality. Particularly in speech quality assessment it is necessary to critically analyze every reactive behavior a human makes. It is also important to look for rules so that the data collected for the voice and speech quality assessment can be validated. The development and use of procedures in quality assessment and the interpretation of their results are always in danger of either oversimplifying the voice and speech perception procedure and thus establishing behavioral norms, or of getting lost in individual behavioral characteristics.

5.4 Speech quality assessment as a measuring process

Naturally voice and speech quality assessment wishes to supply utilizable results. If we ignore the medical diagnosis, the main interest lies not in judgments of individuals, but rather of groups. Therefore an inquiry into quality assessment in the field of technology tends to concentrate on methods involving typical inter-individual features that can be proved. The variety of features of the individual object of voice and speech perception and assessment is limited to those that are really of interest in a comparison, namely those typical features that refer to qualities (in the sense of content) and quantities. In general quantities are determined by drawing up or defining the dimensions. In this respect speech quality assessment can be seen as a comprehensive measuring process in which setting a scale is part of the speech perception and assessment process.

A scale is a means of communication. In auditory speech quality measurements a scale is a standardized surface construct or means of expression that has in-depth, structurally relevant information concerning the speech quality event. Borg/ Staufenbiel [31] talk of empirical relatives and of numerical relatives, i.e. groups of numbers with the corresponding relations and/or operations as the ones perceived.

Scales are constructed according to what information they should transport. A scale is therefore a numerical analog to a description of features of perceived speech quality events, whereby the following should not be ignored. The type and form of a scale should be designed in such as way that the numerical relational system forms an accurate copy of the structures and features of the speech that has been perceived and judged. To put it simply, scales are distinctive relational systems divided into classes, orders or intervals. They are used to represent physical, psychological or perceptive phenomena in static, usable values. Scales are chosen best when the loss of information that occurs in the scaling process (e.g. translating what has been perceived in speech into this kind of relational system) is small compared to the benefit to the process [263]. In this case the allocation has to be such

"[...] that there are analogous relationships between the allocated numbers to certain relationships of object characteristics. In order to be able to deal with individual cases using this method, a group of suitable numbers is sought so that if to the features of the group, or part of the group, numbers were to be assigned, every number would represent an element of the group. This group of numbers is called a scale, the elements of the scale are measured values for the features of the basic elements." [18]

The field of science that generally deals with measurements and scaling is metrology. By definition, this science deals with physical measurands. As such it also deals with basic acoustic topics. In contrast to metrology, voice and speech quality assessment is not exclusively directed towards acoustic but to acoustic-auditory events. It does not deal with physical but psycho-physical, i.e. perceptive features of an acoustic object. Anyhow, metrology provides extremely useful impulses when it comes to qualitatively and quantitatively describing speech events. This becomes clear when we look at the methods of inquiry used in classical psycho-physics from a metrological point of view [267].

Classical psycho-physics has a strong link to voice and speech quality assessment. In general it researches the relationships between physical events and the objects of perception associated with these events. The most important questions posed in psycho-physics are:

• Which signals from the environment are processed by which senses of perception?
• How can percepts be described?
• What conclusions can be drawn from what has been described in terms of signals and signal processing processes?

In the speech quality assessment context the process of speech perception is the prerequisite for the assessment. If speech perception is understood to be a special case in auditory perception (which is part of psycho-acoustics), then it is clear that psycho-physics can provide extremely useful information on how to approach voice and speech quality assessment. However, the used methods and approaches specialize too much on particular aspects of auditory perception to be able to analyze the speech *quality* assessment approach. Psycho-physics and speech quality assessment do have things in common, particularly when the matter is describing individual auditory events. When measured, these individual events are shown in such a way that their standardized format can easily be quantitatively and qualitatively documented, without losing any of their meaningfulness. In the following approaches and methods from psycho-physics and metrology are only referred to when they promote the comprehension and search for solutions in speech quality assessment.

5.5 Measuring and measurands

As mentioned above, measuring is the term used when a physical dimension is to be measured:

DIN Def. »measuring«
   "Carrying out planned activities to make a quantitative comparison of the measurand and a unit." [46]

DIN Def. »measurand«
   "A physical dimension that can be measured." [46]

Measuring refers to the process of quantitatively comparing a physical dimension with a unit. The value of the measurand (which is a physical dimension according to the definition) is expressed by the product of the numerical value and the unit. This means that the value of the measurand is on the ratio level.

Not only physics has the task of measuring. Psycho-physics measures the relationship between physical phenomena and phenomena of perception. The qualitative and quantitative characteristics of phenomena of perception, and therefore also of speech perception events, are, in contrast to physical phenomena, not only scaled on the ratio level, but also at the nominal, ordinal or interval level.

If the definition of dimension in [46] is followed to the letter, it is not appropriate to speak of a measurand when scaling perceptual events, because it is a perceptive and not a physical dimension. At best the term may be used generally if the measurand of the events is expressed by the product of the numerical value and the unit, in other words when the events are scaled on the ratio level. Moreover it would also be incorrect to speak of a »psychological or perceptive measurand«.

Although metrology is an extremely helpful scientific aid in our context generally, these narrow constraints are more of a hindrance here. Therefore in the following the term »measurand« will be used differently in the following sense:

»measurand«
   Feature of an object to be measured which can numerically be described in the course of the measuring process.

We deduce from this definition that the value of the measurand does not necessarily have to be expressed as the product of the numerical value and the unit, i.e. on a ratio level. This is illustrated by the following example that is related to the measurable physical object »acoustic event« and the perceptive-auditory object »auditory event«:

There are various ways to measure characteristics of the physical object »sound«. For example, as to the »strength« of a sound, the sound pressure can be determined. The sound pressure, p, is a physical quantity, i.e. it describes the product of a numerical value and a unit – the internationally agreed unit here being the "Pascal (Pa)".

For physical quantities which are measured on a ratio scale, i.e. by determining the ratio of the measured quantity and a unit of the same dimension, the following holds:

$$a = (a)\,[a]$$ with a ... physical quantity
(a) ... numerical value
$[a]$ ... unit

In the case of the sound pressure such an expression could, for example, read as

$$p_0 = 20 \times 10^{-6}\text{Pa} = 20\mu\text{Pa}$$

where this particular example denotes a specific reference sound pressure, p_0, frequently used in acoustics. Ratio scales have an absolute zero, which means that the quantity vanishes for numerical values of zero.

Besides ratio scales, interval scales are sometimes used for physical quantities. An example in this context is the scale of the sound pressure level, L, where a sound pressure, p, is normalized to a reference sound pressure, p_0, and the ratio obtained this way is then taken to the logarithm as follows:

$$L = 20\log(p/p_0)\text{dB}$$

The letter symbol dB stands for Decibel, which is not really a unit but a unit-like indicator. There is no absolute zero as can be shown as follows. If the sound pressure p is of the same value as p_0, their ratio becomes one and we obtain the result

$$L = 20\log(1) = 0\text{dB}$$

although the sound pressure did not vanish. The sound pressure level scale is an interval scale, where equal intervals in dB denote equal sound-pressure ratios. For example, an interval of

$$\Delta L = 20\text{dB}$$

anywhere on the sound pressure level scale is equivalent to a sound-pressure ratio of 10, since

$$\Delta L = 20\log(10) = 20\text{dB}$$

Characteristics of the perceptual object »auditory event« are measured in a similar way as are characteristics of physical object, though the measurements are performed through human listeners and not by physical instruments. These listeners are instructed to assign numbers to their percepts following predescribed rules. If the assignment of numbers is accomplished by ratio judgment, e.g. by a comparison of the strength of an auditory event to a unit-percept of the same kind, the resulting perceptual quantity is also described by the product of a numerical value and a unit – just like with physical quantities. The strength of auditory events is called their »loudness« and consequently, the internationally agreed unit for loudness, the »sone«, is a genuine unit, the loudness scale is a ratio scale and, as such, has an absolute zero-point.

Yet, there is another scale for the strength on an auditory event, called loudness level. This scale has no real unit but a unit-like indicator called »phon«. The human listeners, when using this scale, do not perform ratio-judgments, but judge on an ordinal scale, i.e. they compare two auditory events and determine which of the two is louder (or softer). Thus they judge on the rank order of the two auditory events with regard to their respective perceptual strengths. Such an ordinal scale does neither employ an absolute zero-point nor do the same intervals of scale values denote equidistant strength intervals (or strength ratios) of the auditory events.

Nevertheless, be it a ratio scale, an interval scale or an ordinal scale, in each case numerical values are assigned to objects such as to reflect relations between the object in a quantitative way. This, however, is the essence of measurement – which holds for physical objects as well as for perceptual ones. Accordingly, we use the terms »measuring« and »scaling« the following way:

In general the aim of both measuring and scaling is to represent the value of a dimension of an object in numbers. In this respect measuring and scaling have the same function. However, the terms are not synonymous:

»measuring«
The entirety of all the activities in the measurement chain up to determining the value of a dimension.

This comprises, for example, the choice of dimension with regards to the object, the measuring equipment, practical application of the measuring principles, determining to what extent expectations or requirements are fulfilled etc.

»scaling«
The entirety of all the activities that are concretely applied to the process of assigning a value to a dimension, which vehicle is the measuring object, on a corresponding scale value (measuring value) according to set rules.

Scaling is therefore a sub-process of the whole measuring process.

5.6 Measuring processes vs. investigative processes

The measuring process is carried out by applying investigative processes. We speak of a measuring process when a special measuring principle and measuring method are practically applied [46]. If the measuring principle and method are not specified, then the process is termed an investigative process. In an investigative process general distinctions are made between

> "[...] assessment, observing, measuring, calculating, statistical estimating processes, or a combination of the above. What is determined is an assessment, observation, a measurement [...], calculation or a combination of the above. Depending on which kind of investigative process is used, the result of the investigation is termed an assessed, observed, measured, calculated or statistically estimated result." [51]

It also makes sense here to change the term because, according to this definition, the measuring process is one of several processes used to investigate the measuring object and scale values. As the term measuring extends beyond the physical to the psychological-perceptive area, it seems pertinent – providing a special measuring principle and measuring method can be practically applied – to take the above definition of investigative process where it refers to measuring process, and to change it as follows:

»**measuring process**«

> A measuring process is an observing, assessing or instrumental process, a calculating, statistical estimating process, or a combination of the above. What is determined is an assessment, observation, an instrumental measurement, calculation or a combination of these. Depending on which kind of measuring process is used, the result of the measurement is termed an assessed, observed, instrumentally registered, calculated or statistically estimated result.

By analogy with the distinction between measuring process and measuring principle the following condition can be attached to the term »instrumental measuring process«: The process is deemed instrumental if a special measuring principle and special measuring method are practically applied. If the »measuring principle« and »method« are not specified, the term preferred is »instrumental investigative process«.

For the measuring process this means that perceptive-auditory phenomena are measured by applying observing processes. Assessing perceptive-auditory phenomena is done by observing and assessing processes, while physical phenomena are dealt with in instrumental measuring processes. Furthermore, calculating processes and/or statistical estimating processes are employed to obtain calculated and/or statistically estimated results. In addition various measuring processes can be combined with each other. All the aspects mentioned refer to investigative processes, measuring processes, operations and results.

5.7 Measuring instruments and measuring organs

The tool or medium for scaling physical phenomena is, as in natural and engineering sciences, termed a »measuring instrument«. The tool (or medium) for scaling perceptive-auditory phenomena is a »measuring organ«. The measuring organ is the perceiving organism, in our context here it is the human listener (animals or other organisms can serve as measuring organisms, depending on the task to be carried out in the measurement). If scaling is used generally, in other words if no distinction is made between physically measurable and perceptively measurable, the measuring medium will in the following be termed the »measuring apparatus«. A measuring apparatus can therefore be a piece of equipment or an organ (e.g. a test subject).

5.7.1 Instrumental measuring methods (a digression)

This book aims to name the conditions and methods of auditory voice and speech quality measuring procedures, and to examine their validity and reliability. Auditory measuring procedures fall into the category of observation and assessment processes, and a thorough discussion of these methods would be incomplete without at least a short discourse to introduce also instrumental measurement methods.

In voice and speech quality measurements there are numerous approaches that aim to calculate or predict the speech quality on the basis of physical measured values [9], [14], [98], [99], [102], [194], [212], [261], [264]. These are described as instrumental measuring and calculation processes, some are quality prediction models. Also included are instrumental intelligibility measures such as the Speech Transmission Index STI [62], [240], and the Speech Intelligibility Index SII [3]. They are based on the transmission function or the frequency dependent signal-to-noise ratio or the modulation-transmission function of the transmission channel. These values are then taken to calculate the speech intelligibility.

Other instrumental methods exist for dealing with noise and temporal distortions, e.g. the Room-Acoustic-Speech-Transmission-Index RASTI [62] or the Speech-Interference-Level SIL [157]. These estimated values are useful in person-to-person communication, public announcements or personal communication systems. They have their limitations, particularly for non-linear systems such as speech codecs with low bit rates (< 16kbit/s), for frequency dependent reverberation times and for impulse background noises.

If a high enough degree of speech intelligibility over a communication channel is realized, a measurement of instrumental intelligibility still does not provide enough information about the overall quality of the system. Particularly for narrowband telephone nets measurements have been developed that should make it possible to give a good prediction of listeners' reactions to the overall quality and the listening effort. In general there are three distinctive types [179], [183].

Class 1: Signal-based comparative models

Measurements in this class are used as a tool for testing new codecs and comparing their performances to others whose performances are already known. They compare the incoming and outgoing signals and predict one-way voice transmission quality. Their measurands are e.g. effects of waveform and non-waveform codecs, background noise and transmission errors. Typical methods are the Perceptual Speech Quality Measure, PESQ [121], PSQM [120], TOSQA [14], and further psychoacoustically motivated measures [100], [102], [212].

Class 2: Network planning models

These models estimate speech transmission quality within the framework of a network plan for a communication system. Predictions are based on planning characteristics of networks which are – in general – available even before the system has been set up. They take into consideration the whole transmission path from the speaker's mouth to the listener's ear, including the linear transmission paths, nonlinear coders and background noise. Plans are currently being realized to enhance existing approaches for time-variant distortions as well as for wide-band speech transmission and different commercial appliances terminals [214]. The most well-known models are the so-called E-model [116] and SUBMOD/CATNAP [122].

Class 3: Monitoring models

Measuring models are tools for monitoring flow in network operation. Ratings support to localize problems. The so-called Running-Quality-Index should enable the user to choose a connection with a predefined quality value. Well-known approaches are the INMD (in-service, non-intrusive measurement device), or the 'Call Clarity Index' CCI [114], [117], [118]. In principle useful quality values can only be obtained if the model they are based on functions adequately. Ideally this model should not only allow speech communication quality to be planned but also to be supervised.

The instrumental measuring and calculating methods are heavily criticized by experts on scaling. Borg/Staufenbiel [31] consider them to offer obscure perspectives in which it is important to formulate calculations and predictions according to given laws and using constants from collections of formulas, whereby understanding the formulas is secondary and only the predictions have to be correct. They justify their opinion with remarks like the following:

> "What is fatal about this situation is that it is ultimately neither completely clear which field a scale value or index is illustrating, nor under which circumstances this index has to function [...] the external validation cannot change this fact, especially as the criterion is mostly just as vague." [31]

Although their remark is extremely critical, they also see something positive in the process and approaches:

"Despite our main doubt, scaling, as an index, is irreplaceable [...] because we can hardly wait for basic research to provide us with something more differentiated. Furthermore, it is feasible that successfully functioning indices can give access to greater knowledge." [31]

This once again reinforces the necessity of having a scientific dispute on the basics of voice and speech quality measuring, as envisaged by this book. The focal point of interest, but also of criticism, does not concern instrumental and calculating processes, but observation and assessment processes of quality measurements. As the results of observation and assessment processes were, almost without exception, used to develop and optimize instrumental measuring processes, this work will have lasting effects on instrumental processes. The aim, as mentioned above, is to make a contribution to basic research and thus to offer voice and speech quality measurements something differentiated. At this point both types of processes (instrumental and auditory) are to be compared:

As has already been said, developing and optimizing instrumental processes is based on questioning listeners who then judge speech quality events. The data related to the listeners are then used to test the performance of processes that have not been optimized, and then, where necessary, to carry out the optimization. If, after a range of system performance tests and system optimization, the listeners' assessments agree with the instrumentally measured data on quality, the instrumental method in question is considered to be valid, or at least promising.

However, often before the real research and development approaches have been completed, these instrumental methods are rashly over-generalized and categorized as objective processes for measuring voice and speech quality. This happens because in the planning and development phases of different speech technology products there is a great demand for fast and economical measurements of quality. This leads to cases, like the following, where the speech quality of speech transmission technologies is the indicator of the quality of the systems under investigation. The speech transmission systems should have the best possible design and therefore they are optimized in an iterative process of speech quality measurements and systems' revisions. As the relevant auditory observation and assessment measurement processes are both costly and time-consuming, it is understandable that instrumental measuring processes are in great demand. The choice of process is clearly guided by what benefits the qualities will bring, which is partially expressed in the fact that purposefulness tends to have the upper hand over accuracy.

This general approach is well-known and, within limits, understandable. What is dubious though, is when instrumental processes are described as »objective measuring processes«. This name is misleading and tends to suppress the beneficial characteristics of instrumental processes. This does not only apply to laymen, but also to experts in voice and speech quality measurements. The impression is given that the processes are able to totally model the perceptive and assessment behavior of a person with normal hearing capabilities. This may be true for a simulation of a listener's assessment behavior in a communication situation, but does not apply to the listener's previous experience, in other words for a prognosis on how a listener will assess a future speech sound event. A listener's behavior in new communication situations is rarely transparent and comprehensible, so it is extremely difficult to predict.

The inconsistency and apparent unpredictability of the assessment behavior is an argument taken up by some supporters of objective speech quality measuring processes. They claim that objective processes would produce better measurement results than a person making assessments, because the instrumental process would deal with invariance and not with the spontaneous behavior of the individual. Moreover, behavioral aspects have no place in the speech quality measurements.

This author does not agree with this view. By developing instrumental processes, an attempt is made to model the perceptive and assessment behavior of a person with normal hearing abilities. One part of characterizing human behavior must take into account that when appraising something, a human is not always constant.

However, this does not mean that data obtained using test persons rather than instrumental processes make it possible to predict future speech quality events. Even when conceiving observation and assessment processes (e.g. listening tests), an artificial scenario is designed in the attempt to measure the speech quality of spoken utterances. Then, using these values, predictions on future quality events are made. Even here there are people who claim that their measurements are done »correctly« because they are working with the measuring apparatus »man«.

This view can also be refuted because both the application of observation and assessment processes and instrumental measurements of speech quality require that the situation on which the experts focus, where speech occurs acoustically, be defined as accurately as possible. The potential influences should be identified and all this should be taken into account in the choice of process, when specifying the tests and interpreting the results. This requirement has, of course, no chance of being fulfilled. As a result, neither of the processes (observing and assessing nor instrumental), are convincing when it comes to predicting quality events.

5.8 Summary

The issues discussed above within the scope of perception theories produces the following results:

• The terms »perception«, »perception event« and »percept« must be clearly differentiated: Perception is a process, the perception event is something original linked to the senses, and a percept is the result of reflection within the perceptive world of a perceiving individual.

• The perception event is an experience (phenomenal). As such it is not communicable.

• What is perceived is a description (phenomenological). As such it is the intellectualized form of a corresponding perception event.

• Describing and communicating what has been perceived can only be done with a sign system. The description is also phenomenological.

• Perception is not absolute, but refers to relationships. What is perceived is the result of a comprehensive perception process in a special situation.

• Not all the signals that are directed towards the senses are processed. Only those are picked out that are, at that moment, possible vehicles of information for the perceiver.

• The perception process obviously proceeds strategically, e.g. by concentrating on invariable phenomena. Special perception and cognition strategies for conceptualizing what has been perceived are used. There are many indications that this can be both intra- and inter-individual.

• Isomorphs can be observed when describing what has been perceived. These isomorphs cannot be compared with reaction norms. Deviations from expected reactions are not atypical. They are an expression of individual behavior that ultimately sets the standard. Individuals can react atypically. The aim is to understand what has triggered off this behavior.

• Just because many of the individual descriptions of perceptions, expectations or assessments of objects appear to be similar, this does not mean that we can conclude that whatever is behind the signs is really what should have been transported.

• Quality events are the products of different mental prerequisites and influences that control the mental process.

• Despite all the individual and variable aspects of the assessment process, there is always a certain degree of inter-subjectivity aimed at the speech quality measurement (with knowledge of the intra-subjective parts).

• A »measuring instrument« is a tool or medium for scaling physical phenomena

• The tool (or medium) for scaling perceptive-auditory phenomena is a »measuring organ«. The measuring organ is the perceiving organism, in our context here it is the human listener

• If scaling is used generally, in other words if no distinction is made between physically measurable and perceptively measurable, the measuring medium will in the following be termed the» measuring apparatus«. A measuring apparatus can therefore be a piece of equipment or an organ (e.g. a test subject).

6 Quality Assessment in View of System Theory

Instrumental measurements as well as observation and assessment measurements of speech quality can be characterized by different physical and psycho-physiological functions and their corresponding scaling functions. The implications of these measuring and scaling processes become clear when speech quality measurements are viewed as a theoretical system.

The starting point for this has been taken from Blauert's approach (cf. [15], [16]), illustrating functions of sets in connection with consciously perceiving systems (in other words measuring organs). This approach will be briefly presented here, so that it can be extended and applied later.

According to Blauert, perceiving systems can be modeled on the basis of the following fundamental sets (cf. Fig. 6.1.):

- the set of acoustic events S_0 with the elements s_0
- the acoustic scale S with the elements s
- the set of auditory events A_0 with the elements a_0
- the auditory scale A with the elements a
- the set of descriptions D_0 with the elements d_0
- the descriptive scale D with the elements d

Developers of instrumental measuring processes aim at constructing measuring equipment that illustrates the function $a_0 = f\{s_0\}$. Such a measuring apparatus should allocate an element s from the acoustic scale S to the acoustic element s_0, that is equivalent to the measuring value d as an element of the descriptive scale D, d being the scaled form of the description d_0 of the auditory event a_0 (which once again correlates with the acoustic event s_0). The desired function is thus:

$$s = f\{s_0\} \text{ and } d = f\{s_0\} => s \Leftrightarrow d$$

When applying this approach to voice and speech quality assessment one aspect comes to the fore which Blauert neglected, namely the fact that the function $a = f\{a_0\}$ does not exist; a_0 cannot be scaled directly because scaling is a reflective process. Also, it has already been mentioned that the function determining $d = f\{s_0\}$ (in other words the scaled description d of the acoustic event s_0 correlated with this scaled result) cannot be seen as a fixed process. This is made clear once again when the individual steps in the judgment process are listed:

The aim of measurements is to determine the function $a_0 = f\{s_0\}$. But because the auditory event a_0 can only be communicated as a description d_0, a_0 cannot be

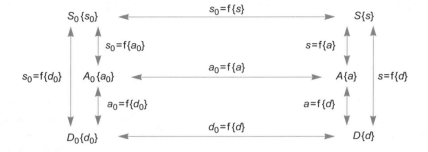

Legend:
- the set of acoustic events S_0 with the elements s_0
- the acoustic scale S with the elements s
- the set of auditory events A_0 with the elements a_0
- the auditory scale A with the elements a
- the set of descriptions D_0 with the elements d_0
- the descriptive scale D with the elements d

Fig. 6.1. Relationships between fundamental sets and scales in auditory experiments according to Blauert [18]

scaled. This means that $a = f\{a_0\}$ cannot be determined. The auditory event a_0 can only be communicated via its description d_0. The description d_0 is, in turn, communicated by d, an element of the descriptive scale D.

In the observation measuring process a_0 is assigned the description d_0 – by applying the conventions to the description of a_0 (thus $d_0 = f\{a_0\}$), it becomes what is perceived, reflected and to be communicated. The relationship $d = f\{d_0\}$ makes it possible to directly apply the measured value d, as a function of d_0.

If the quality of the acoustic event s_0 or the auditory event a_0 is now to be assessed, the following set and scaling functions can be identified for instrumental and assessment measuring processes for speech quality assessment:

- the set of the acoustic events S_0 with the elements s_0
- the acoustic scale S with the elements s
- the auditory events A_0 with the elements a_0
- the set of what has been reflectively perceived Q_0 with the elements q_0
- the set of the descriptions D_0 with the elements d_0
- the descriptive scale D with the elements d

In contrast to the system shown above, in speech quality measurements the set of »quality events« is added to the set of »auditory events«. An auditory event a_0 becomes a quality event q_0 (by drawing attention to the desired and perceived quality), which like a_0 is only communicable as a description d_0, i.e. just as a_0 also q_0 cannot be directly scaled and the function $q = f\{q_0\}$ cannot be analyzed. The quality event q_0 is only directly communicable via the description d_0, whereby between q_0 and d_0 it holds true that $d_0 = f\{q_0\}$.

The measuring chain in the quality assessment process of s_0 or a_0 looks like this:

$$a_0 = f\{s_0\}, \ q_0 = f\{a_0\}, \ d_0 = f\{q_0\}, \ d = f\{d_0\}$$

In these functions it holds true that d as an element of the descriptive scale D is a quality measuring value $d = f\{s_0\}$ (observation and assessment processes) related to the object s_0, and s as an element of the acoustic scale S is a measuring value $s = f\{s_0\}$ related to the object s_0, and d as an element of the descriptive scale D correlates with s as an element of the acoustic scale S. The processes are depicted in Fig. 6.2.

This list of the various measuring and scaling functions shows that the quality judgment of speech goes through different phases of transformation until it is eventually allocated a measured value on the descriptive scale. When the proportion of

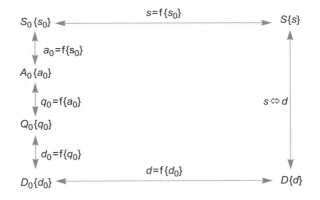

Legend:
- the set of acoustic events S_0 with the elements s_0
- the acoustic scale S with the elements s
- the set of auditory events A_0 with the elements a_0
- the set of quality events Q_0 with the elements q_0
- the set of descriptions D_0 with the elements d_0
- the descriptive scale D with the elements d

Fig. 6.2. Functions of sets in connection with scaling quality events

the reflection on the speech that has been perceived during the measuring and scaling processes is relatively small, the correct terminology is »conversion«. However, when the degree of intellectualization for what has been perceived in speech dominates in the process of determining the measured value, we talk of »translation«. This is of course an artificial division, because it is only possible to estimate a proportion of the reflection on what has been perceived. When listener-related measurements and scales of acoustic events are being performed, there is never a real conversion process, because a sign system always has to be applied to communicate what has been perceived in speech ($d = f\{a_0\}$). On the other hand translation does not mean that the measuring value allocated to an auditory event cannot arise from highly unpredictable individual behavior and is therefore not a value. Translating individual behavior becomes an inter-individual object of comparison because conventions are applied to the description of what has been perceived in speech, the judgment and communication of the speech quality assessment.

In view of these considerations it makes sense to regard conversion and translation as the theoretical poles of a continual scale, whereby individual measuring and scaling tasks can themselves be scaled by their degree of conventionality. According to its definition quality assessment is reflective. What is heard is compared to what is expected, weighed up and a decision is made. Once this has been done, the results are applied to a previously defined scale. The whole process always leads to a loss of information.

7 Auditory Measuring Procedures

7.1 Measuring and scaling methods in psycho-acoustics

As has already been discussed, psycho-physics examines the relationships between particular physical phenomena and the perceptive events associated with them. Psycho-acoustics is a branch of psycho-physics that investigates the relationships between auditory events and sound. Wundt and Fechner are the acknowledged founders of classical psycho-physics [68]. Their aim was to mathematically describe the character, size and intensity of phenomena that have been perceived [29]. Apart from the classical observation and assessment methods, their means of detecting the absolute threshold, the differential threshold and points of subjective equality have become standard methods:

Detecting the absolute threshold (L):
Exemplary questions: What maximum value can the sound pressure of a sinusoidal sound x have for the listener not to be able to perceive the auditory event? Or: What minimum value can the sound pressure of a sinusoidal sound x have for the listener to just perceive the auditory event?

Detecting the differential threshold (JND = Just Noticeable Difference):
Exemplary questions: If a tone x is given, what value must the sound pressure of a comparable tone y have for the listener to perceive a difference? Or: What value must the sound pressure of a comparable tone y have for the listener not to perceive any difference?

Detecting equivalents (PSE=Point of Subjective Equality):
Exemplary question: Given a tonal sound x with a particular relational frequency (e.g. 1kHz) and a defined sound pressure level, what value must the sound pressure of a comparable tone y of a relational frequency have for the listener to perceive both tones x and y as having the same pitch?

The measured values in these three methods (also called method of limits) can be determined using the method of adjustment and the method of constant stimuli:

Method of adjustment:
The stimulus is adjusted until a predetermined condition is fulfilled. One example of the method of adjustment is given by the usual way of measuring via a tone au-

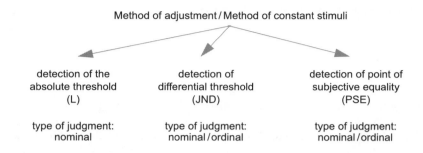

Method of adjustment / Method of constant stimuli

detection of the absolute threshold (L)	detection of differential threshold (JND)	detection of point of subjective equality (PSE)
type of judgment: nominal	type of judgment: nominal / ordinal	type of judgment: nominal / ordinal

Fig. 7.1. Method of adjustment and method of constant stimuli

diogram. The test supervisor (or the test person) gradually increases the sound pressure of a sinusoidal signal until the test person indicates that he/she can hear something. This is repeated for different frequencies. The stimulus is modified and presented as often as necessary until the condition of perceptibility is fulfilled.

Method of constant stimuli:

In this method the most suitable judgment is chosen from a battery of offered judgments. The meaningfulness of the results is largely influenced by the initial choice of the stimuli in the preparation phase of auditory tests. For example, care must be taken that the predetermined range of stimuli has the desired degree of discreteness of auditory phenomena with respect to the goal of the assessment. Inevitably this means that a measurement using this method can only be as good as the stimulus material.

In psycho-acoustics assessment measurements are used to discover, among other things, the relationship between attributes of acoustic signals that can be measured physically (e.g. sound pressure) and the accompanying features of auditory objects (e.g. loudness). Accordingly, with regard to speech the aim is not only to measure the acoustic phenomenon »speech signal« and/or the auditory phenomenon »speech sound«, but also to define the relationship between these two, by means of measurements. The results of the measurements serve to extend knowledge of the relation between speech signal and speech sound and then to define it.

As the investigation into this relationship has the highest priority, every effort is made to choose measuring objects (acoustic signals or sounds) that minimize potential errors in the measurement. Errors can be caused by complex characteristics of the object to be measured. If the measurements solely aim at determining the function and not the characteristics of signal and sound, »simple« signals (e.g. white noise or sinusoidal tones) are often chosen as an input to a measuring instrument. If the same »simple« signals form the input to the auditory system, the measuring organ only perceives »simple« sounds as an object in most cases. The instrumental and auditory measured values are compared and contrasted once the signal (instrumental measurement) and the sound (auditory measurement) have been scaled. The aim of the measurement procedures is to reduce the degree of conventionality within the measuring and scaling processes to a minimum so that the listener's opinion can be seen as a transformation rather than a translation process (as to the differentiation between transformation and translation see page 73).

However, in psycho-acoustics there are many different objects to be investigated (measuring objects) that are neither physically nor perceptibly »simple« (in the sense that they result in a one dimensional measured value). Speech quality is one such case.

Speech quality is a multi-dimensional phenomenon. Accordingly one of the main aims of measuring is first to identify all the measurands by means of observation and/or assessment (analysis of dimensions), whose carrier the measuring object »speech« is. The next step is to determine procedures to measure the values of the measurands that have been found. Only when all the measurands have been identified, it is possible to examine the functions common to speech sounds in more detail, by comprehensively measuring speech quality. However, if single measurands are highlighted, for example as part of an experiment, this produces measured values that can be attributed to the formation of classes, rankings and possibly even the distances of these measurands, but that cannot make general statements concerning speech quality. Thus one question remains unanswered – if an answer is at all possible – namely how the results of measurements of the individual measurands fit into an overall value of the measuring object »speech quality«.

When specifying measurements one of the tasks involves ensuring that the number and type of measurands and the composition of their features are adequate for the nature of the measuring object. This is one condition that is particularly pertinent to speech quality assessment. Comparing the values acquired from instrumental measurements and those from observation and assessment measurements to obtain speech quality makes this patently obvious:

In physical acoustics, measurands such as sound pressure, sound pressure levels, fundamental frequency, or signal duration are typically distinguished. For every measurand (e.g. sound pressure level) there are measuring apparatus (e.g. sound level meters) that, depending on the input quantity to the measuring apparatus (e.g. time-dependent sound pressure), assign measurands (e.g. a equivalent continuous sound pressure level) to the measuring object (e.g. a speech sequence).

Auditory acoustics is similar: We know that comprehension, naturalness, speaker recognition, clarity or speech velocity are measurands of spoken utterances. Here too, measuring apparatus (e.g. test persons) are at our disposal for different measurands (e.g. comprehension) that, depending on the input quantity of the measuring apparatus (e.g. single words), assign measurands to the measuring object (e.g. a speech sequence).

A comparison of the measurands and measured values of physical and auditory measurements shows that they are related to different elements or features of the measuring object (as to the differentiation between quality elements and quality features see page 16). If the functions common to the acoustic speech signal and the auditory speech sound are to be determined, it must first be ensured that the results of the instrumental and auditory measurements are in fact comparable; in other words that they measure corresponding aspects. A level of comparison must exist that allows the individual measurands of the »speech signal« and the ones of the »speech sound« to be related to each other. This involves operationalizing, which throws up the following question: How can measurands whose carriers are a measuring object be identified and made so precise that their values can be measured by measuring instruments and organs, and that the acoustic and auditory values can be directly or indirectly compared to each other? Or using an example: How do we measure the comprehension of spoken utterances in such a way that the instrumental processes used on the one hand, and the assessing measurements (the auditory process) on the other hand, result in »correct values« with respect to the measurand »comprehension«, so that this can then be used as a basis on which to determine the function of comprehension common to the acoustic speech signal and the auditory speech sound? In this context it is necessary to introduce the term »correct value«:

DIN Def. »correct value«

"A known value for purposes of comparison, whose deviation from the true value [...] can be regarded as negligible for the purposes of comparison." [46]

This example illustrates a fundamental difference between the tasks of classical and speech related psycho-acoustics: In classical psycho-acoustics objects are commonly measured using known, clearly defined measurands in order to discover values and to draw conclusions on the relationship that exists between acoustic speech signal and auditory speech sound based on the listener's reaction. However, in speech-related psycho-acoustics the physical measurands related to the speech signal may be known, but there is no classification of these measurands − or of the features of speech sounds in general − in systematic categories that determine the speech quality event. Therefore speech quality measurements have to consider the following question:

Which approach will produce a systematic classification of the object »acoustic speech signal« that leads to speech quality events which can be measured perceptively and, if necessary, physically?

Speech quality measurements do not solely seek to measure values of spoken utterances, but also to take measurements that will result in a classification of the object »speech« itself. Quality features and elements that are measurands of speech quality must be identified. A taxonomy has to be established that endeavors to adequately describe the features of the speech quality event in terms of segments and classes, and to weight each feature in terms of its contributing to the overall speech quality. In other words all the measurands of the measuring object »speech« have to be analyzed in a systematic way.

This means that speech quality measurements can pursue very different goals. One goal is to determine the values of single measurands. The measurands are found by analyzing measuring objects, and they can, in turn, be used for similar measuring objects (if features of similarity can be defined). Another goal is to examine whether supposed features really are measurands of a particular measuring object.

Whereas the quality of the measurand is not called into question in the first case (as the intention is solely to ascertain the measuring data), in the latter case the measurement serves the purpose of analyzing the measurands and checking whether they qualify (functional aspect of measuring data).

The varying aims of measurements obviously affect the choice of principle and method. The emphasis is first placed on the assessment principle and method, in particular on the process whereby measured values of single measurands are recorded. In this context the method of adjustment and the method of constant stimuli can be used to detect the absolute threshold, analyze the differential threshold and equivalents in order to be able to measure individual features of the auditory speech event (i.e. of the measuring object):

> Regarding the loudness of auditory events, for example, investigations are carried out to find out whether such an object is audible or not (L), whether it is slightly louder than the reference object (JND) or whether it is perceived to be as loud as a reference object (PSE).

Using these methods, it is possible to measure special aspects of the qualitative and, to a lesser extent, the quantitative features of the object »speech sound«. Under »qualitative feature« it is meant that the result of the scaling solely applies to a class feature, while »quantitative features« analyze their absolute and relative characteristics. The listed methods are applicable for contexts in which analyzing the class and ordinal characteristics have top priority, but they cannot cater for expansions or distances, i.e. the relative and absolute characteristics.

Consequently both the method of adjustment and the method of constant stimuli deliver results on the nominal and/or ordinal level. They can effectively be applied to special aspects of speech quality assessment. However, when a more comprehensive assessment is required, they have to be extended by methods on the interval and ratio levels. In the following we give two examples of such methods: The method for determining perceptual differences is one of the measuring methods that involve interval scaling, the method involving magnitude scaling is in the ratio scale group:

Method for determining perceptual differences (additive intervals):
Exemplary question: If we assume we have the speech sounds x_1 and x_2, we choose the speech sound y so that it lies in terms of perception (e.g. regarding the feature »loudness«) in the exact middle between x_1 and x_2 (method of halving). Using this

method the perceptual differences are determined by the position of the point y on the continuum between x_1 and x_2 (perceptual differences).

Magnitude estimation:

Exemplary question: Given the speech sound x, choose the speech sound y so that it is exactly twice as loud as x (method of doubling/perceptual relationships)

Interval and ratio scaling are not only used to quantitatively analyze class and ranking characteristics, but also for relative and absolute relations.

7.2 Measuring scales

In auditory quality measuring scales serve to log and communicate the results of auditory observation and assessment measurements (as for speech quality observation methods are normally not used, they will be skipped in the following). In general scaling attempts to quantify measuring objects. Scales have a surface form, which, on the one hand is the postscript of judgments, but, on the other hand, also constitutes the regulations on how to treat the measured values. A scale gains acceptance and is judged to be more user-friendly when its »code« (in the sense as a finite repertoire of clear signs) can be used intuitively, or when there can be no misconceptions on how the sign from one repertoire (of the thought and the judgment) can be applied to the signs of another repertoire (of the scale).

The scale is required to be of the same dimensions as the measurand. The form should allow the composition and characteristics of judgments to be numerically represented and should also enable classifications, orders and rankings to be presented when the relevant scales are applied. Thus in speech quality measurements with test persons, the scale has to be designed to be the notation of content and structural patterns in judgments relevant to speech quality.

7.2.1 Standard scales

But what does such a scale look like? In natural science and technology different scales are used for physical phenomena which are ultimately based on the basic dimensions of length, mass, time, electric current intensity, thermodynamic temperature, amount of substance and luminous intensity. For each of these basic dimensions that are not related to others, there is a unit (meter, kilogram, second, ampere, Kelvin, mol, candela) [86]. Physical scaling signifies how often a unit fits into a dimension. The format of such a scale (e.g. a scale for measuring length) is given in advance and is standardized.

When measuring perceptual phenomena, there are also standard scales that have been determined beforehand (cf. Fig. 7.2.). However, they cannot be used for all measuring objects. One reason for this is that there are not always defined basic dimensions and unitary systems for all the measuring objects, as is the case with physical measuring. From all this the conclusion can be drawn that it is desirable to make the metrology of the area of application »speech quality measurements« to the object of research, if all the facets of the field are to be taken into account. This has already happened in certain fields.

One good example can be found in research into pain where new measurement principles, methods and processes are being researched and developed to augment existing knowledge of »psychological and neural mechanisms of pain and their evaluation« [61], [211]. In speech quality measurements this is at present little more than wishful thinking. Individual issues may be examined in greater detail, but a global approach which would allow the results of individual issues to be evaluated and integrated with others does not exist. In practice this means that only a few issues of speech quality measurements can be considered without compromises or concessions.

Listening-quality scale:

Quality of speech	Score
Excellent	5
Good	4
Fair	3
Poor	2
Bad	1

Listening-effort scale:

Effort required to understand the meanings of sentences	Score
Complete relaxation possible; no effort required	5
Attention necessary; no appreciable effort required	4
Moderate effort required	3
Considerable effort required	2
No meaning understood with any feasible effort	1

Loudness-preference scale:

Loudness preference	Score
Much louder than preferred	5
Louder than preferred	4
Preferred	3
Quieter than preferred	2
Much quieter than preferred	1

Fig. 7.2. Opinion scales acc. to ITU-T [119]

7.2.2 The scale as a function of the measurand

In order to be able to measure the perceptual phenomena that correlate with speech quality, the characteristic to be measured in each case has to be determined in the planning and specification phases. Only when the measurand is known a corresponding scale can be specified. This detection process of the measurand and method largely determines, along with other factors, the quality of the measurements. Thus, finding out the measurand (the perceptual dimension) and the process to measure the value of the dimension constitutes the elements of quality of the measurements. This again points at the task of operationalization.

In the context of speech quality measurements operationalization is related not only to perceptive phenomena, i.e. determining the quality features of speech sounds relevant to the listener, but also to physical phenomena (e.g. taking into consideration the known elements of quality of the speech signal). Quality features are analyzed with respect to the measurement aims.

Initially the individual measurands which largely determine speech quality do not have to be known. This is particularly the case for exceptional measuring objects, for example »synthetic speech«. In order to be able to design the most appropriate measuring scale, it is essential that the measurands of speech quality are systematically analyzed. Different approaches are available to master this task, selected ones will be introduced below.

One approach involves initiating a pilot project on dimension analysis to discover which events of perception and quality occur in the listener, which characteristic signals and which features the listener notices. Schulte-Fortkamp's process of intermittent thinking aloud can be used as an example [227]. Ideally test persons are asked to describe their auditory (or even multi-sensory) and quality event in free associations, their perception and assessments being guided as little as possible. The descriptions are then analyzed and channeled step by step so that they lead to precise attributes of the auditory object being assessed. If the number of test persons is large it becomes possible to discover important features/attributes of what has been perceived and assessed, and fundamental measurands can be identified. In this context must also be mentioned the so-called semantic differential which is one of the most common methods of analyzing dimensions.

If the measurands that dominate the assessment processes are known, the format of the measuring scale can be established. The scale serves to numerically depict the perceived characteristics of measurands.

So far scales have only been mentioned in conjunction with different scale properties. However, scale properties are, in principle, only related to the aspect of transformability of scale classes. If scale properties are agreed upon, concrete scales have to be defined. Every scale class can therefore have various scale representations specific to its particular area. The individual scale representatives are first designed in the pragmatic context of a measurement (e.g. depending on the age structure of the test group). Assessment tests are used to scale the expected values of quantities of certain quality features or attributes of speech sounds. These quantities are commonly described verbally. Each measurand identified in a preliminary test (e.g. comprehension, clarity, acceptance) is then scaled, i.e. classified according to the categorical descriptions that quantify the auditory object. But first a check is normally carried out to see if already existing standard scales like the »listening-quality scale«, the »listening effort scale« or the »loudness-preference scale« are eventually appropriate verbal descriptions of typical quantities of the detected measurands. Verbal descriptions are the best choice when test persons under similar test conditions all have to put each acoustic stimulus into the same category of scales, however often the task is repeated. In other words they react in the same way. Moreover, discrete characteristics of the object under investigation can also be listed without a loss of information occurring, with the result that all the communicable features of the speech perception and assessment event have been recorded. As has already been shown in Fig. 7.2., the »listening-quality scale« has the following categories:

The individual categories are accorded numerical values between 5 and 1 (or 1 and 5), which serve to verbalize and quantify the object of perception and assessment. Such a scale is often used wrongly, because when analyzing the results many assume that they are looking at numerical relatives scaled in intervals which will permit parametric tests to be carried out. Brownen/McManus [36] for example have examined and analyzed different category scales. The test persons were given the task of putting terms related to scales from all the known category tests into a rising sequence. The test was done both for American-English and Italian and led to the following result:

"It has been shown that category scales do not possess interval properties, and that therefore parametrical statistical tests (which require normal distributions) are not appropriate. Generally, between-point judgements are not allowed (i.e. no. 2.5 or 4.5 answers), and observers avoid the end points or boundaries of scales, further reducing the amount of information which can be gathered. It is now clear that there is very little sensitivity built into these scales. It follows then, that if information other than just rank is sought, more sensitive scaling methods should be used. Such measurement methods as ratio or graphic scales yield information about distances between stimuli which other measurement methods do not." [36]

There are comprehensive examinations of verbal and numerical labels of points on a scale and concerning the question of how many points are necessary to scale a sensory event (cf. for instance [28], [36], [58], [59], [76], [171], [241], [262]). Amongst other things, the aim of these investigations is to eliminate the above mentioned disadvantage of category scales: They do have ordinal properties, but category tests can also have interval properties. However, it should not be assumed that every category test has interval properties. Each scale level is calculated according to the »law of categorical judgment«. Using this process variance can be calculated, which in turn, defines the distance between the intervals on the scale. The width and the

0	Nothing at all
0.5	Extremely weak (just noticeable)
1	Very weak
2	Weak (light)
3	Moderate
4	Somewhat strong
5	Strong (heavy)
6	
7	Very strong
8	
9	
10	Extremely strong (almost max)
•	Maximal

Fig. 7.3. Borg's CR-10 category-ratio scale [27], [30]

boundaries of each category have to be defined (for more detail, see [1], [92], [232]). One example to extend classical category scales to ones with interval and even with ratio properties is Borg's approach [27]. Unfortunately, this approach has not been universally accepted in speech-related psycho-acoustics so far (cf. Fig. 7.3.).

In summary: It has been shown that the definition of scales is a major task for speech quality measurements. Standard scales do not always have to be the best solution for a given task. The approach in which category scales labelled for one language are simply translated word for word into another language is extremely dubious. Here an analysis of this particular semantic field is called for to identify equidistant terms appropriate for interval properties of category scales. This is an issue that linguistics needs to address.

The tests described here provide a number of different measurands whose individual quantities can be categorized and quantified by means of selected verbal descriptions coupled with numerical relatives. At the end of the tests the data must be analyzed. Using suitable statistical procedures (e.g. correlation analysis) correlations between the individual attributes are calculated and metrically represented. In addition processes analyzing dimensions, such as a factor analysis, enable us to investigate which main components or dimensions of the auditory speech perception and judgment even dominate, which measurands are complex values, i.e. that are composed of measured values from different dimensions, and which speech perception and judgment events can be scaled uni-dimensionally. The results of such a data analysis are used to measure objects and form the foundation for specifying further measurements on speech quality.

The description of an exemplary process for analyzing quality features and their verbal and numerical relatives within the framework of a task to measure and scale auditory speech perception and judgment events makes it clear how much effort is

Table 7.1 Quality features of natural and synthetic speech

natural speech, transmitted over telephone	synthetic speech
acceptance	intelligibility
overall impression	general quality
listening effort	naturalness
comprehension problems	precision of articulation
articulation	accuracy of pronunciation
pronunciation	pleasantness of voice
speaking style	adequacy of word stress
voice pleasantness	appropriateness of tempo
	liveliness
	fluency

required to achieve reasonable measurements. Particularly when the aim is to measure a speech perception event quantitatively, it is extremely important to have appropriate measurands and scales so that the most important characteristics and their values can be shown.

When measuring speech quality the individual measurands of the object of perception and judgment are often unknown. Therefore previously defined rules on measuring and existing scales that are not specifically for this object can be a hazard. It is possible that the object is falsely transformed or translated, thus annulling the results of the measurements. Methods that isolate the measuring object and measure physical quantities have a decisive influence on the result of the measurement and the theoretical formulas [253].

This problem becomes apparent when synthetic speech is measured. The measuring object »synthetic speech« has many things in common with the measuring object »natural speech«, but not enough to warrant using established measurement procedures from the field of quality measurements of natural speech without further thought. This becomes clear in Table 7.1 when contrasting the most important quality-defining attributes of natural and synthetic speech according to the ITU-T [117] and Jongenburger/Van Bezooijen [139].

Jongenburger/Van Bezooijen do not only use a different set of attributes but also different numerical scales: The ITU recommends a 2 point scale for the first attribute and a 5 point one for the rest, while Jongenburger/Van Bezooijen prefer a 10 point scale throughout.

With regards to speech quality measurements the steps described above for determining measurands and constructing scales are at the same time the preparations for the measurements. Whether optimized scales are only used for a certain object under investigation or for others too, depends on the degree to which the speech quality elements or the perceived speech quality features deviate from object to object. The more similar the quality elements of perceived quality features are, the greater the probability that the scales can be used without first modifying them. However, the similarity of perceived quality features can, on the other hand, only be determined by measurements.

The choice of scale and its construction are also important contributions to the design of speech quality measurements. Here the responsibility of the test supervisors is clear: They must be able to judge what the measurement actually hopes to achieve. If they ignore the questions posed above, they will make a bad judgment of something that is potentially very promising [163]. The persons in charge of the measurements are expected to know which quantities influence the results of the measurements, how to check them and how to assess their importance. But speech quality is a measuring object that refers by definition to the acoustic-auditory matter alone (in other words to form characteristics of speech), which the listener tends to process as a speech sign, i.e. the carrier of information. As soon as the meaning of what has been spoken accompanies or even dominates the measuring or assessing of speech quality, cognition, experience, context and others begin to play a decisive role, and thus have to be included in the measuring plan. This obviously overlaps with other disciplines such as communication theory or semiotics. Their inclusion, however, strengthens the relatively vague connection between speech perception and speech comprehension in quality measurements – which is not the topic of this discussion here.

Despite of all these different facets of speech quality measurements, speech quality has to be measured using means that are available here and now. The multitude of theoretical considerations may help to recognize influential quantities in measurements, but should not lead to the speech quality measurement being abandoned. In contrast: We should learn from this that the results of measuring speech quality cannot be given too much significance, but they should be recognized as having special properties that characterize the whole measurement.

7.3 Summary

The most important fundamental issues are:

• Can the person in charge of the planning and specification phases of speech quality measurements unveil all the quality features of speech and describe them in such a way that all the quality characteristics and their corresponding measured values can potentially be represented by numerical relatives?

• Are there appropriate scales for all the quality features of speech that have been found, so that all the perceived characteristics of the measuring object/measurand can be numerically represented?

• Are the assumptions on the quality features of speech made by the experiment supervisor prior to the measurement also relevant when for persons who actually run the experiment (question of generalization)?

8 Formal aspects of speech quality measurements

In previous chapters different individual aspects and criteria were mentioned that color the results of measurements of speech quality. If one considers the measurement to be an entity of quality, then each aspect can of it be seen as an »element of quality of measurements«. This means:

A measurement only attains quality when the quality of its elements is designed in the most effective way to suit its goal. The more effectively the influence quantities of measurements are detected, checked and scaled, the more meaningful are the results of the measurements. It might even be possible to achieve the best possible design if, within a cost-benefit analysis, a measurement of relatively lower quality leads to a higher quality of usability. Planning, specifying and carrying out measurements as well as evaluating their results are often individual phases in the design process of speech quality measurements. This effects the usefulness of the results of the measurements: Speech quality measurement results do not generally have any absolute value, but are always to a certain extent relative and specific.

This perspective leads to a critical view on standardized speech quality measuring processes. They may have the decided advantage that they can be carried out under predetermined conditions in different places and their results compared with each other. However, they have one great disadvantage: Results of measurements can be values of a measurand which is not particularly an important feature of the corresponding measuring object (validity). This is specifically the case for the measuring object »speech quality«.

8.1 The measuring object »speech«

In the context of speech quality assessments, speech is the measuring object. The object »speech« exists as an expression of movement; as such it is the product of communicative action. It is an object linked to its use. Hermann talks of speech as an event that is dependent on »the whole cognitive store of individuals« [107]. As a consequence, speech is measured in view of its usage. If concrete measuring objects of speech have to be determined for measurements (e.g. test stimuli), then conversational situations – or, more generally, practical areas in which speech happens as a function of – cannot be ignored [169]. But what are the typical characteristics of speech in that line of thinking?

In order to get closer to answering this question we first draw on the measuring object »synthetic speech«. Synthetic speech is an object that, with few exceptions, essentially is generated independent of its use. Synthesizers can, in principle, be applied in a multitude of practical activities where speech happens, however they have so far proved inadaptable to the acoustic-auditory »form requirements« going along with these contexts. The system input (e.g. written text) is simply transformed into synthetic speech signals by a machine, irrespective of the meaning it is to convey and irrespective of concrete functionalities in real applications. Although there are approaches to distinguish between certain types of functions and the corresponding types of forms, they cannot be linked algorithmically so far.

A wide range of conversational situations can be perceived as being impaired if, for example, the quality of synthetic speech is measured against the expected quality of natural speech – which form is functionally bound to these situations. The characteristics of natural speech are linked to its communicative function, and as synthetic speech is perceived similar to natural speech, the perceived quality of synthetic speech cannot be separated from its functional connection to typical communicative situations.

This aspect was already considered in Chapter 4 within the discussion of developing synthesis. However, it is worth mentioning once again because it goes hand-in-hand with the far-reaching requirements of measurements that in particular refer to determine the measuring object:

In order to be able to compose suitable representative speech stimuli as measuring objects, the communicative processes in typical situations must first be documented and systemized. Only when this systematic foundation has been laid we can be sure that all the measuring objects possess those essential characteristics on which each form is based. As part of a different experiment, Lüschow calls for a general theory of practical activity, which can easily be applied to the task being discussed here:

"Against this background has to be taken a theoretic phonetic/linguistic approach which has not been »pragmatically changed« at a later stage, but has been tied into a general theory of practical activity." [169]

In our context that means: To successfully design speech quality measurements it is necessary to know the possible areas of application, the contexts of speech events. This ensures that all the chosen measuring objects are representing specific speech activity patterns which suitably depict both the typical characteristics and the variety of the corresponding functional entity (i.e. the possible area of use). It is extremely risky to »pragmatically change« at a later date the results of the measurements that have been extracted independent from their application context (cf. [70], [154], [251]).The aim is therefore to make use of the advantages so that speech quality measurements can be not only standardized but also flexibly designed.

By typifying the elements of quality of measurements, we get closer to an analysis of the speech quality. The starting point when planning a measurement does not necessarily involve choosing the best from the range of known speech quality measurements, but, bearing in mind the aim of the measurement, to plan and specify the measurement in detail. Only then should the availability of standard methods be examined [78]. Analyzing the surrounding conditions can make it useful to carry out a standardized measurement, and if necessary to select a compromise solution, but it can also be highly advisable to conceive a totally new measurement. Irrespective of which solution these considerations lead to, the first steps are always to specify and plan the speech quality measurements. Measurements are tools which cannot be applied uniformly, and these tools are oriented towards the characteristic measurands to be defined. These characteristic measurands are derived from a formal consideration of speech quality measurements in general.

In order to attain these aims voice and speech quality assessment offers a variety of measuring procedures to find and assess features that determine quality. The most important aids refer to processes that:

• control and if necessary guide the attention of the person perceiving the speech
• reflect the perceived speech
• assess the perceived speech
• communicate the perceived speech quality event

Aspects of speech quality measurements are not only related to perception, assessment and communication processes, but may also be derived from further fields that are concerned with preparing and concluding these processes, such as:

• choice of vocabulary (linguistic-pragmatic aspect)
• choice of speech material (acoustic-phonetic aspect)
• choice of listeners
• deciding the question format for the test persons (abbreviated TP hereafter)
• deciding how test stimuli should be presented
• analyzing the data obtained
• interpreting the results

In other words: The objects of speech quality assessments are speech sounds that the test subjects perceive auditorily and that they qualitatively and/or quantitatively assess in a certain situation. The object of assessment does not necessarily have to be the speech sounds themselves (e.g. when assessing the quality of voice generation systems). Other measurands that color the speech quality event may come to the fore.

Ultimately the entire sound field, i.e. all kinds of speech signals and noise belong to the scope of speech quality measurements. The test is the means of assessing speech quality. The term »speech quality test« is defined here as follows:

»speech test«
    A routine procedure for examining one or more empirically restrictive quality features of perceived speech with the aim of making a quantitative statement on these features.

Following the essential features on tests, not every experiment for diagnostic purposes can be deemed a test but only those that fulfil the following criteria [163]: A
test

- is scientifically substantiated
- can be routinely carried out under standard conditions
- allows a relative definition of position of the object to be investigated with respect
  to other objects or a certain criterion
- checks certain empirically restrictive characteristics, behavioral dispositions,
  abilities, skills or knowledge

In previous chapters numerous aspects of speech quality measurements and system
assessment were examined from totally different angles. Mostly it was enough to
simply point out that certain aspects or features of the measurement or environment
influence the results of the measurements and that it is essentially necessary to control these contexts. Anyhow the question as to how they influence behavior and results and how they are to be controlled was usually omitted. There are several
reasons for this: The most important one being that knowledge of correlations with
different events in speech quality measurements often refers to special single events
and cannot simply be transferred to the rest of the experiment. The quality of speech
occurs in many different contexts and is unique each time because the speech and
listening situations that color the speech quality event are always different. In general it can be said that almost every speech quality measurement, especially in a
communicative context, is a new task with several variables. The greater the number
of unknown variables there are, the more difficult it is to perform the measurement
and the more likely it is that the measurement will become an estimation.

In speech quality measurements and assessments there are different measuring
methodologies and methods, as well as procedures for testing and evaluating. They
cannot all be introduced here. However, there are given examples which produce a
greater insight into the topic and allow to see what possibilities and limitations there
are in the common procedures for measuring speech quality used today.

8.2 Toward the quality of speech quality measurements

In general it can be said that when speech quality tests are examined, the focus is
placed either on analyzing the tests and their factors of influence (the test as the real
purpose) or on the test as an instrument for measuring speech quality (test as a
means to an end). In this chapter the former will be looked at in greater detail: The
background, prerequisites of measurements for assessing speech quality will be analyzed. This discussion serves to help to understand, plan and check test procedures,
and it is an important contribution to legitimizing measuring methods and under-
mining the ways processes are conducted. The main emphasis is put on the aspects
of »quality of speech quality tests«, starting off with the following question: What

experiments and approaches are there to ensure the quality of auditory speech quality measurements? There are comparatively few publications dealing with specific aspects of quality of speech quality measurements. The most important contributions are listed below:

- on scaling: Categorical Estimation, Magnitude Estimation, Paired Comparison, Reaction Time Measurement ([8], [19], [29], [43], [44], [60], [83], [138], [166], [195], [196], [197], [266], [268])
- on recording the individual differences in perceiving synthetic speech ([103], [217], [231])
- on finding out the dimensions of listeners ([112])
- on the role of the internal reference system when assessing speech stimuli ([219])
- on the effects of training and attention ([87], [220], [228])
- on the perceptibility of speech sounds in speech quality tests ([176], [209])
- on the influence of contextual effects ([12], [73], [83], [84], [85], [162], [172])
- on the role of cognition in speech quality tests ([167], [44], [45], [26])
- on drawing attention and its control ([191])
- on the influence of inter-individual speaker characteristics (native speaker/non-native speaker) within speech quality assessment ([193])
- on the influence of interference at the speaker's or listener's end on results of measurements ([237], [25], [135])
- on the meaningfulness of results of measurements ([230], [82], [81], [129], [130])
- on the influence of test material on the results of measurements ([207], [206], [77], [141])

Many of the experiments in the literature cited above are called »tests« although it is often debatable whether the processes actually meet the requirements of tests. As individual special effects are rarely analyzed in connection with speech quality measurements, these »tests« should perhaps be termed »studies« or »experiments«. Examples of investigations that deal directly with extending knowledge of speech quality measurements and their respective factors of influence are listed in Table 8.1.

Table 8.1 Speech quality measurements: factors of influence

objective	reliability of speech intelligibility tests
measuring procedure	'comprehension tests'
measuring object	synthetic speech
measurand	speech intelligibility
carrier	consonants of CVC pattern
results acc. to Van Bezooijen[248]	test results can be reproduced under test/re-test conditions; reliability proven

Table 8.1 Speech quality measurements: factors of influence

objective	inter-individual variety in view of different measuring procedures
measuring procedure	SAM_SUS Test
measuring object	synthetic speech
measurand	speech intelligibility
carrier	semantic bearing monosyllables, embedded in a non-semantic bearing sentence
results acc. to Hazan/Shi [103]	the relative difference between worst and best test results, averaged over all stimuli, is about 28%; all TPs can be characterized as a relative homogenous group
objective	inter-individual variety in view of different measuring procedures
measuring procedure	SAM CVC-identification test
measuring object	synthetic speech
measurand	identification of consonants
carrier	consonants of CVC pattern
results acc. to Hazan/Shi [103]	the relative difference between worst and best test results, averaged over all stimuli, is about 47%; possible explanation: global and analytic listeners; all TPs can be characterized as a relative homogenous group
objective	age of TPs as a factor of influence of a segmental intelligibility test
measuring procedure	SAM CVC-identification test
measuring object	synthetic speech
measurand	identification of single consonants
carrier	consonants of CVC pattern
results acc. to Howard-Jones [112]	results seem to be dependent on age; more data necessary to prove hypothesis
objective	acquaintance with synthetic speech as a factor of influence of a segmental intelligibility test
measuring procedure	SAM CVC-identification test, open response mode
measuring object	synthetic speech
measurand	identification of single consonants
carrier	consonants of CVC pattern
results acc. to Howard-Jones [112]	TPs who are acquainted with synthetic speech gain better results (79% for unexperienced, 62% for experienced listeners)
objective	acquaintance with synthetic speech as a factor of influence of a segmental intelligibility test
measuring procedure	unguided listening
measuring object	synthetic speech
measurand	sentences
results acc. to Boogard/Silverman [26]	in an unguided situation the listener adapts to the peculiarities of synthetic speech; the adaptation time is dependent on the synthesis principle

Table 8.1 Speech quality measurements: factors of influence

objective	influence of the synthesis entities on the adaptation ability of TPs
measuring procedure	SAM CVC-identification test, open response mode
measuring object	synthetic signals generated by an allophone and a diphone synthesis system
measurand	identification of single consonants
carrier	consonants of CVC pattern
results acc. to Jongenburger/Van Bezooijen [139]	data indicate that the adaptation time is shorter for allophone synthesis than it is for diphone synthesis. Possible explanation: allophone synthesis is rule-based, and construction principles become easy to see through so that listeners are able use them for recognizing
objective	comprehension of narrative speech
measuring procedure	comprehension test for adults
measuring object	synthetic speech, natural speech and silence reading
measurand	aspects of comprehension
carrier	15 narrative passages
results acc. to Schwab et al. [228]	learning effect for comprehension of synthetic speech; after a short period of time no significant difference between the 3 different measuring objects

The aim of these studies is to examine individual aspects of measurements in order to recognize uncertainties and to compensate for systematic bias errors. This must be done if results of measurements are to be adequately evaluated. The top priority is therefore to understand the conditions and factors influencing measurements.

For reasons of economy methods for measuring synthetic speech are often used that have already been proved to be suitable for natural speech. Although synthetic and natural speech signals have much in common (they are both carriers of information) the form of the synthetic signal differs greatly from its natural counterpart. The use of such processes for the measuring object »synthetic speech« and the analysis of collected data often calls into question the meaningfulness of such results.

If the measurements of artificial speech are to have quality, then it goes without saying that features being unusual in comparison with natural speech should be analyzed and the influence of single events on the results of the measurements be investigated. To this end there are carried out investigations that deal with individual parameters of the measurement scenario, namely whether any influence, and if so exactly which influence, do certain parameters of the measurement scenario have on the results of the measurements. Ideally the focus is placed on one particular aspect while attempting to keep all the other influential factors constant. The parameter of interest is then measured by repeating the phases of variation of the parameter and analyzing the listener's behavior.

When finding out the measured value of single parameters of speech quality, it often comes to light that the methods are tailor-made for the question on hand. This means that the methods used to examine factors of influence are often specially designed tools that do not generally serve to measure speech quality or specific aspects thereof, but which contribute to an increase in the quality of speech quality measurements. Once again only when each of the tools is the best possible for this purpose, can the results of these experiments increase knowledge of the subject.

As was mentioned above, even today such parameters usually become the objects of investigation by chance rather than by design, particularly as the resources to examine such parameters are limited. As a consequence, the probability of succeeding in gathering the best possible selection still lies in the realms of chance, intuition and experience rather than knowledge of the subject.

In summarizing the scientific considerations of speech quality tests, it can be said that in knowledge there are still enormous gaps that must be closed. However, nobody should draw the conclusion that available results of speech quality measurements are totally useless. Without doubt not all the events and processes of every measurement have been analyzed scientifically, but this does not mean that none of the measurements and results have quality. The mere fact that speech quality experts find themselves drawn more and more into the quality loop of speech technology and telematics shows that speech quality measurements are of great benefit. However, this should not mislead anybody into believing that there is no urgent need to identify constitutive parameters of the whole measurement process to make it more transparent and thus easier to control. This should be done not only for interest in science, but also for pragmatic reasons.

Even if it cannot be expected that speech quality measurements in all their facets will be scientifically investigated and conclusions drawn, this discussion shows that simply because there is no scientifically proven paradigm of all the essential elements of quality of speech quality tests, adequate descriptive and classification systems do help to attribute the results of the measurements to the most suitable benefits.

8.3 Speech quality tests as measuring tools

Auditory measurements have the purpose of judging and scaling the perceived quality of speech events. Measuring is the means or tool to determine the value of a dimension. It is a tool to put information on auditory and quality events into a quantitative form which again is the basis for data analysis and interpretation.

In general a tool is an entity that was formed for a special purpose and with which help something else is formed or produced. A tool is therefore an aid that contributes to completing a process reducing duration and effort needed to reach the goal. The quality of a tool is therefore recognized if after its use:

- the target has been achieved
- in reaching the target time has been saved
- the effort of reaching the target has been reduced

The quality of the tool is derived from its use. When this refers to the tool »speech quality test« this means that the quality is always a function of its use in the test run. If established tests are used according to their definition and rules, quality can be attributed to them. This presupposes that there are generally recognized and binding rules (e.g. norms, standard regulations) that give instructions on the use of such tools when applied to speech quality tests. In other words: Speech quality tests require a certain degree of standardization.

In the field of speech quality measurements there are a number of standard test procedures that deal with scaling different aspects of perceived speech quality. These tests can be used to measure intelligibility, prosody, overall quality, hearing impairments, speech impairments, to analyze speech competence or to judge auditory-verbal communicative competence. In addition speech tests are also tools in diagnostics and therapy.

8.3.1 Intelligibility tests in speech technology and audiology

On looking more closely at the different measurands for which speech quality tests are done, it becomes obvious that the naming of the measurand (e.g. intelligibility) is not enough to select the best possible test from the range of available standard processes. In particular in speech quality measurements there is often more than one measuring process available (if it is at all described in such general terms as above). This can be seen with the example of speech intelligibility tests. The best known processes are listed in the following and described in more detail with reference to their test vocabulary and purpose. For phonemic notation the SAM Phonetic Alphabet is used here. This is a standard phonetic alphabet, like IPA (cf. [72]). Phonemes are symbolized as '|', phones as '/'. Within speech technology, the following processes for speech intelligibility belong to the standard procedures:

SAM Standard Segmental Test   Howard-Jones[111]
Test vocabulary: Single words with the structure CV (consonant–vowel), VCV (vowel–consonant–vowel), VC (vowel–consonant) that mostly do not carry any meaning, a 'C' that stands for all the possible consonants of the target language, 'V' for the vowels /a/, /i/, /u/ only. The vocabulary contains stimuli as:
/ta/, /ti/, /tu/, /pa/, /pi/, /pu/,... /ata/, /iti/, /utu/, /apa/, /ipi/,... /at/, /it/, /ut/,...
The test material is available for English, German, Swedish and Dutch.
Task: Please write down what you have heard. No restrictions on possible answers. Computer-based analysis of the results.
Aim of the measurement: to analyze how well single consonants are identified in initial, medial and final positions of a word.

Cluster Identification Test (CLID)   Jekosch[126]
Test vocabulary: non-semantic bearing monosyllables which are correct with regard to phonemotactics, having the structure C_nV, C_nVC_n, VC_n. Items are generated by computer depending on the objective of the experiment (usually 300 stimuli). Statistical information is used to instruct the computer which stimuli it should present: /StRi:/,.../kRaUSt/,.../?Ytst/,...
Task: please make a note of what you have heard. No restrictions on possible answers. Computer-based analysis of the results.
Aim of the measurement: to analyze whether consonant clusters in monosyllables can be identified.

Bellcore Test   Spiegel et al. [238]
Test vocabulary: 312 C_nVC_n items. Monosyllables are phonemically correct and each type consists of a meaningful and a meaningless token: swan–swog,... warmth–dorth,... dropped–globbed,...
Task: Please make a note of what you have heard. No restrictions on possible answers. Computer-based analysis of the results.
Aim of the measurement: to analyze how well groups of consonants in monosyllabic words can be identified.

Diagnostic Rhyme Test (DRT)   Fairbanks [80], Voiers [252], Peckels/Rossi [198]
Test vocabulary: 192 meaningful groups of monosyllables with the structure CVC. Articulation of the initial consonant is analyzed. In each case six words are formed which differ from each other by means of six phonetic contrasts of the initial consonant (voicing, nasality, sustension, sibilation, graveness, compactness). Pairs of stimuli are formed: veal–feel, reed–deed, vee–bee, sing–thing, weed–reed, key–tea, ...
Task: Listener hears one word at a time and has to mark on an answering sheet which one of the two words he/she thinks is correct. Restricted choice of answers. Computer-based analysis of the results, averaging the error rates.
Aim of the measurement: to analyze how well groups of consonants in monosyllabic words can be identified.

Modified Rhyme Test (MRT)   House et al. [110], Logan [164], Sotscheck [236]
As an example, German test vocabulary: 900 groups of semantic bearing monosyllables with the structure CVC. The articulation of both the initial and final consonant as well as the vowel is analyzed. In addition to the acoustical stimulus word, five other rhyming words are formed (initial, medial, final) that are presented to the TP visually. From the group of six words on offer the TPs make a note of what they have heard:
Acoustic stimulus: /haUs/
Visual response paradigm: <raus>, <Maus>, <Laus>, <Haus>, <saus>, <Klaus>.
(The test for the English language consists of 50 ensembles of 6 monosyllabic words which makes a total set of 300 words).

Task: Please mark the word you have heard. Restricted choice of answers. Computer-based analysis of the results.
Aim of the measurement: to analyze how well consonants and vowels in monosyllabic words can be identified.

Harvard Psychoacoustic Sentences   Allen et al. [2]
Test vocabulary: a closed set of 100 semantic bearing sentences chosen according to frequency of occurrence of individual phonemes in the English language.
<Rice is often served in round bowls.>
<Glue the sheet at the dark blue background.>
<These days a chicken leg is a rare dish.>
Task: Write down what you have heard. Computer-based analysis of the results.
Aim of the measurement: to analyze word intelligibility in a sentence context.

Haskins Sentences   Allen et al. [2], Nye & Gaitenby [192]
Test vocabulary: a closed set of 100 non-semantic bearing sentences chosen according to frequency of occurrence of individual phonemes in the English language.
<The great car met the milk.>
<The old corn cost the blood.>
<The short arm sent the cow.>
Task: Write down what you have heard. Computer-based analysis of the results.
Aim of the measurement: to analyze word intelligibility in a sentence context.

Semantically Unpredictable Sentences (SUS)   Benoit et al. [13]
Test vocabulary: 50 semantically unpredictable sentences that consist of meaningful monosyllabic words: <The hair stands on the light tooth.>
Five types of sentences are represented, whereby each sentence consists of approximately seven chunks.
Task: Write down what you have heard. Computer-based analysis of the results.
Aim of the measurement: to analyze word intelligibility in a sentence context.

This list of standard speech intelligibility tests in speech technology shows that even a vague characterization of the test vocabulary reveals great differences in the individual test procedures. These differences are due to the specific requirements and conditions of the different applications. These tests were originally developed to solve certain questions (e.g. Sotscheck's rhyme test was designed to test intelligibility of speech transmission systems in telecommunications) and were then applied to comparable problems.

The differences in the speech intelligibility tests even increase if we leave speech technology and look at those tests that are used in audiology (in Germany). In this field characteristic forms of the signals are not examined (as is necessary in speech technology). The tests are carried out to measure a patient's hearing deficiencies. Amongst other things, intelligibility is an indicator of how well the auditory system is working.

Schubert [226], Hahlbrock [96]	Numbers test
Döring/Hamacher [53]	Aachen Logatom Test (ALT)
Döring/Hamacher [53]	Three syllable test
Kalikow et al. [140]	Speech Perception in Noise Test (SPIN-Test)
Tschopp/Ingold [247]	Speech Perception in Noise Test (German version of SPIN)
Bench et al. [11]	Speech in Noise Sentence Test
Hagermann [94], [95]	Closed set Swedish Sentences for Speech Intelligibility Test
Kliem [144], Kliem/Kollmeier [145]	Two syllable rhyme test for German
Nilsson et al. [188]	Hearing in Noise Sentence Test (HINT)
Chilla et al. [41]	Göttingen Speech Intelligibility Test
Kollmeier/Wesselkamp [150]	German Sentence in Noise Test (Göttingen Sentences)
Wagener [255], [256], [257]	Closed Set German Sentence in Noise Test (Oldenburg Sentences)
Brand/Kollmeier [33]	Adaptive versions of the Göttingen and Oldenburg Speech Tests

All the speech tests that have been named so far in this chapter are analytical, diagnostic processes that involve either identifying single speech components (e.g. phones, syllables, words) or concentrate on the listener's ability to identify speech. Although both aspects are scientifically closely connected, there is a difference between 'identifying' and 'being capable of identifying'. This distinction also makes it easier to define 'intelligibility': Intelligibility is related to the function of being able to comprehend statements and having the comprehension to pass on these statements to the recipient. Comprehensibility and comprehension are thus understood in the following way:

> "A statement is said to be comprehensible when it can be reiterated by the recipient in the sense of the sender on a certain level of comprehension." [32]

In audiology comprehensibility tests seek to analyze the ability to identify. The only reason they are mentioned here is to point out the different uses such tests can be put to. In the following emphasis will be placed on the possibility to identify with regard to signal characteristics on offer, i.e. on describing the elements and features of quality of speech signals and speech sounds, whereby an intact auditory system is assumed to be in place. The object to be examined will usually be synthetic speech.

In the case of synthetic and natural speech intelligibility is given the highest priority, but is certainly not the only quality feature of speech sounds. If the speech sounds are intelligible, then other features gain significance, for instance stress, prosody, acceptance, naturalness, effects of interference caused by transmission techniques or the overall quality. The most important processes among those features that have just been named are:

Prosody:

Grice et al. [90], Howard-Jones [111], Sonntag [234], Sonntag and Portele [235]: Prosodic Form and Function Tests

General Quality:

Howard-Jones [111]: SAM Overall Quality Test

ITU-T Overall Quality Test (Category assessment test):

Acceptance (Do you think that this voice could be used for such an information service?):

> 1: yes
> 2: no

Overall impression (How do you rate the quality of the sound of what you have just heard?):

> 1: excellent
> 2: good
> 3: fair
> 4: poor
> 5: bad

Listening effort (How would you describe the effort you were required to make in order to understand the message?):

> 1: complete relaxation possible
> 2: attention necessary, no appreciable effort required
> 3: moderate effort required
> 4: effort required
> 5: no meaning understood with any feasible effort

Comprehension problems (Did you find certain words hard to understand?):

> 1: never
> 2: rarely
> 3: occasionally
> 4: often
> 5: all of the time

Articulation (Were the sounds distinguishable?):

> 1: yes, very clear
> 2: yes, clear enough
> 3: fairly clear
> 4: no, not very clear
> 5: no, not at all

Pronunciation (Did you notice any anomalies in pronunciation?):
       1: no
       2: yes, but not annoying
       3: yes, annoying
       4: yes, very annoying

Speaking rate (What do you think of the average speed of delivery?):
       1: much faster than preferred
       2: faster than preferred
       3: preferred
       4: slower than preferred
       5: much slower than preferred

Voice pleasantness (How would you describe the voice?):
       1: very pleasant
       2: pleasant
       3: fair
       4: unpleasant
       5: very unpleasant

All these tests are tools for measuring perceived characteristics of speech sounds. Speech sounds themselves are usually seen as momentary events who are carriers of the features to be investigated.

This aspect draws attention to the problem of how representative test materials are when the aims of tests are decided upon. The dynamics of speaking (and listening) behavior in communicative situations, for example, have so far been left out of all the standard speech tests that have been discussed here.

The following approaches address speech intelligibility as a function of the ability to form a *dialog*:

Heinrichs [104], Kettler et al. [143]	Kandinski Test
Richards [216]	Postcard Test
Möller [177]	Short Conversation Test

8.4 Summary

• The aim of speech quality assessment is to first identify measurands of the measuring object »speech«.

• Speech quality assessment can be seen as a comprehensive measurement procedure in which scaling is only part of the whole process.

• The aim of both measuring and scaling is to represent the value of a measurand in numbers. A measured value is understood here as a numbered value of every process involved in scaling.

• The term »measuring« is understood to mean all the activities in the whole measuring chain that together determine the value of the measurand. Scaling, on the other hand, refers just to those activities that concretely deal with the process of attributing the quantity to be measured and the scale level and the measured quantity and the scale value. Scaling is one part of the whole measuring process.

• In speech quality measuring scales are numerical relatives that record and communicate the measured results of speech-related observations and assessments.

• The instrument or means of scaling physical phenomena is called the »measuring instrument« here; the medium for scaling perceptive, auditory phenomena the »measuring organ«. A measuring apparatus can therefore be an instrument or organ (e.g. a test subject).

• Decisive parameters of speech quality measurements are type and format of the scales. Irrespective of the theoretical considerations, in every planning and specification phase of measurements the first task is to design concrete scales to carry out the special measurements.

• Each scale category can have different scale representatives. They are designed when the pragmatic context of the measurement has been decided upon. The format of the measuring scale is adapted to suit each operational speech quality object, in other words the features of what has been perceived and assessed.

Amongst other things, the following aspects have an influence on speech quality measurements:
• scaling
• individual differences in the perception of synthetic speech
• the internal reference system when assessing speech stimuli
• training effects
• attention
• perceptibility of sounds
• contextual effects
• cognition
• attention and its control
• individual listener characteristics
• test material

9 Towards the Structure of Speech Quality Measurements

In the previous chapters exemplary processes for analyzing different aspects of voice and speech quality were presented. The stimuli used were both natural and synthesized speech sounds. The measurements were carried out with totally different aims in mind. They were introduced here to give an overall impression of the processes and aims of measurements. The processes were mainly developed for the field of speech technology. In this chapter, however, open questions and aspects that are thrown up by these measurements will be discussed.

As is clear from previous chapters, there are a number of useful studies on specific aspects of speech quality. Many, however, do not go beyond elementary single events because their positions within the field of speech quality measurements cannot be defined. This is because there is no single comprehensive concept for speech quality measuring and no corresponding structure available in which results of measurements could be classified.

This chapter seeks to redress this omission by working out a conceptual structure for speech quality measuring. The basis of this will be the discussions that have so far taken place. All those aspects and components of speech quality will now be systematically examined, i.e. they will be given an order and structure. In this concept the single components will be created and ordered in such a way that any change to one of them will affect speech quality as a whole. There will be named the most important coordinates and their mutual dependencies that are essential for describing the states and processes in speech quality measurements.

9.1 Taxonomy for quality of speech synthesis

Van Bezooijen/Van Heuven [249] also aim to impose a structure upon speech quality measurements, even if their selection encompasses less than the one presented here. They restrict themselves to a taxonomy of methods for measuring the quality of speech synthesis, mainly dependent on the measuring object and the measuring procedures (cf. Fig. 9.1.). Their approach involves a mixture of an analytical process on the one hand and the measuring object »speech synthesis« on the other one. The following measurement criteria have been taken into consideration for analyzing the quality of speech synthesis:

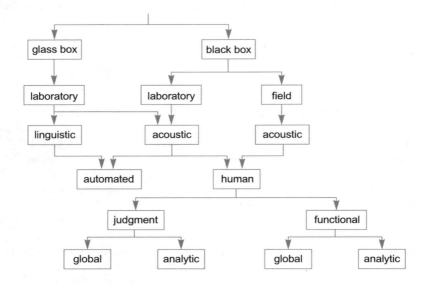

Fig. 9.1. Taxonomy for measuring the quality of synthetic speech acc. to Van Bezooijen/
       Van Heuven [249]

- Should the system itself, or should the quality of the functional sub-entities of the system be investigated (glass box)?
- Should the quality of the whole system in the sense of its effect of the system performance on the listener be investigated (black box)?
- Should the measurements be performed in a laboratory or in a field test?
- Should linguistic or acoustic elements be examined?
- Should observation or assessment measuring procedures (human) or instrumental measuring procedures (automated) be used?
- Should evaluations of features of synthetic speech be given (judgment) or should communicative functions be measured (functional)?
- Should a global evaluation be analyzed or should analytically measured values be taken for the measuring object?

This taxonomy is particularly helpful for specifying and planning speech quality measurements in speech synthesis. Important aspects of the measuring object are raised which make it possible to specially classify measuring methods that refer to the most important features of the object »speech synthesis« being investigated. The most important aspects for a comprehensive structural model of speech quality measurements in general are not recorded, because the aim was not to develop such a model. However, the approach makes clear how much more meaningful measurements become if they are clearly typified and their discerning structural features named. This taxonomy allows the relationship between the analytical process and speech synthesis to be described. The only task that remains is to undertake a functional analysis, for instance to analyze the conditions, aims and decisive criteria.

9.2 Aspects specific to usage

As has already been mentioned in other places, all the discourses in this work stem from the conviction that the quality of entities, whatever they may be, manifests itself only when these entities are put to use. This also applies to the entity »speech quality measurement«. The term »usage« refers to the application of measuring procedures (or to analytical processes generally) and to the interpretation of the measured values obtained. Measurements are carried out for a specific purpose, and the way in which the measurement is performed and the use to which the measured values are put fulfils this purpose. In literature on the subject distinctions are drawn between (cf. [42]):

- evaluation of the system with regard to the applications (evaluation)
- assessing system performance or functions of system components (assessment)
- assessing for diagnostic purposes, or to draw up a performance profile (diagnosis)

Accordingly, these terms are used as follows:

»**Evaluation**«: determination of the fitness of a system for a purpose – will it do what is required, how well, at what cost etc. Typically for a prospective user, may be comparative or not, may require considerable work to identify user's needs.

»**Assessment**«: measurement of system performance with respect to one or more criteria. Typically used to compare like with like, whether two alternative implementations of a technology, or successive generations of the same implementation.

»**Diagnosis**«: production of a system performance profile with respect to some taxonomization of the space of possible inputs. Typically used by system developers, but sometimes offered to end-users as well.

These three measuring motives are clearly distinct and each requires totally different methods to collect the data. However, they all view speech quality measurement as a consumer-service task so that the entity is essentially the »speech sound« as the product.

Speech quality measurements cover far more. In measuring speech it is either the means or the purpose. It is the means when the results of the measurement are directly implemented to refer to another object. It is the purpose when the aim of the measurement is to find out something about speech sounds themselves. The same can be said of processes measuring speech quality. The different functions of both

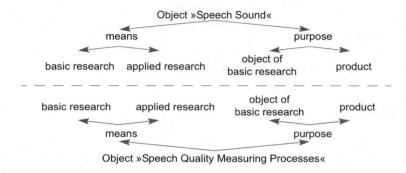

Fig. 9.2. Speech and measuring processes: aspects of use

objects, i.e. the quality of speech sound and measurement procedures for speech sound quality, are summarized as follows:

The object »speech sound«

Means
- basic research (e.g. speech sound to research cortical activities)
- for applied research (e.g. speech sound in the context of developing hearing aids)

Purpose
- object of basic research (e.g. examining speech sound in the context of different speech styles)
- product (e.g. speech sound as the output of a synthesis system)

The object »measurement procedures for speech quality«

Means
- basic research (e.g. procedures for researching perception of quality of sound coloration)
- for applied research (e.g. procedures for developing speech transmission techniques such as »Voice over Internet Protocol«, perception of non-stationary noise)

Purpose
- object of basic research (e.g. investigating influencing factors of measurements)
- product (e.g. processes measuring speech quality as a consumer service)

These aspects of speech sound and measurement procedures for speech quality are illustrated in Fig. 9.2.

9.3 Systemic view of speech quality measurements

A description of all the different speech quality measuring processes reveals numerous differences, as discussed in Chapter 8. A systematic view has the aim of determining the scope, components, structures and correlations of the processes for measuring speech quality. Achieving this aim is made more difficult by the fact that each new measurement that is carried out, irrespective of whether it has been newly designed or simply repeated, is influenced by certain factors that cannot be distinctly captured. This has already been seen when individual issues concerning speech perception and assessment were discussed. This means that not every process of speech quality measurements draw attention to all the factors that can influence measurements. Often the influences cannot be exactly identified and described. Hidden influences come to attention when they show a strong correlation with changes to results of measurements. But even then they do not have to be describable in the sense of the smallest segmental entities, simply because they occur in a complex context.

Therefore a model based analysis of speech quality measurements will surely fail if the approach is based only on the distinctive features as mentioned above. This does not mean that speech quality events cannot be modeled – an elaborated approach is to determine the function of the constitutive elements of speech quality measurements. As has often been said, speech quality measurements are specified according to the purpose they are to be put to. Such elements include the measuring aim, the measurand, the measuring object, generating stimuli (of the vocabulary) and the measurement process. The functions that correspond to each of these entities are determined by orienting the measurement towards the usefulness of the results. Each measurement can have more than one function, but often one function is focused upon and the others are of less importance.

A structural model of speech quality measurement only becomes viable when it can functionally support the relevant components and features [131], [136]. The most important criterion of functionality is the degree of relevance a change in parameters would have for the meaningfulness of any results of a measurement.

In Fig. 9.2. the parameters that are relevant to the functions of speech quality measurements are listed as part of the structural model. In most cases the parameters are independent; only a few are dependent on others (mutual dependency is graphically depicted by connecting lines).

The model aims to make speech quality measurements more transparent. It offers a form to enable criteria to be made more precise, to structure those contexts that support the practical processes in specifying and planning speech quality measurements – depending on which use the model is to be put to – with respect to the measuring objects »speech sound« or »processes measuring speech quality«. In addition it makes it possible to record and document measurements that have al-

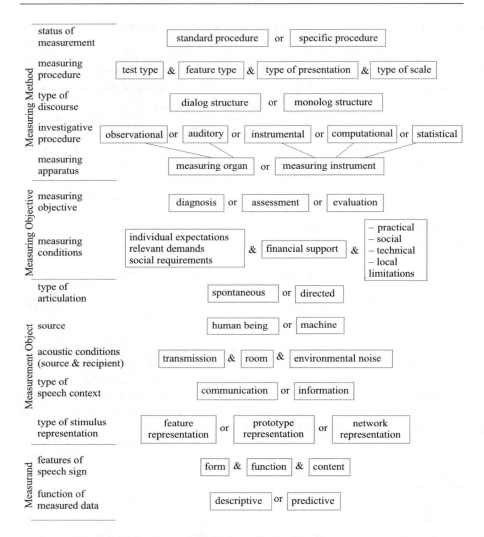

Fig. 9.3. Projection model of voice and speech quality measurements

ready been carried out so that their speech quality or the quality of their components can be analyzed with respect to the chosen paradigms. Thus both speech quality measurements themselves and their position in comparison to other processes and results can be clearly determined. This also eradicates, or at least restricts, the seeming anonymity and randomness of the measurements. The measurements and the results of the measurements become functional. They can be integrated into a global concept, thus creating a basis for dealing with every speech quality measurement in a wider context. This leads to greater knowledge of the subject, and considerably increases the benefit and meaningfulness of the results.

It is not claimed here that this model is complete, but has been designed so that it does not contradict the current situation in science and practice. Where necessary it offers all the possibilities that are presently on offer to enhance it.

10 Segmental Intelligibility: A Dimension of Quality

The goal of this chapter is to systematically consider quality measurements for segmental intelligibility. In the previous chapters, numerous components of speech quality and projections of measurement procedures were presented, discussed and systematized. This chapter follows with an in-depth look at the example of segmental intelligibility as *one* dimension. It shows that – in the mathematical sense – this dimension is in turn a function of several variables, for which speech quality assessment today still has no solvent equation. One reason for this is that certain linguistic questions that determine the results of these measurements still have not been answered and in part have yet to be posed.

The dimension to be discussed in the following is speech comprehensibility and intelligibility. Speech intelligibility is the dominant dimension of speech quality. When speech sounds are not intelligible, communication cannot be successfully accomplished. The expectations of the listener are not fulfilled. No quality is assigned to the speech signal.

Measuring speech intelligibility proves to be a multifaceted task. The leitmotif of this chapter is determining exactly individual quality elements of speech sounds; additional emphasis is placed on the measuring object. Speech intelligibility is restricted to segmental intelligibility – in part, as will be shown, also only to segmental comprehensibility. It is of benefit that numerous procedures to judge speech intelligibility are documented and that important experience in the use of these procedures in quite different measurement contexts is available. This is not necessarily the case for other dimensions of speech quality (e.g. naturalness). The tests measuring segmental speech intelligibility discussed here are further limited to diagnostic methods. Instrumental processes are not taken into consideration. Diagnostic methods are introduced, first of all, and classified according to their most important features.

The focus, therefore, is on the quality of the measuring object »speech«. In this context, a further criteria is discussed, namely the effect of guided and unguided speech perception and assessment in regard to the test results. Both are essential matters of speech metrology that up to now have not been the subject of a systematic analysis.

Thus the theme of this chapter is the scientific analysis of quality elements of speech quality measurements with emphasis on diagnostic methods for segmental speech intelligibility and comprehensibility in the context of pure research. The tests and investigative processes are discussed and analyzed in view of the measuring object speech, whereby aspects of the content of the respective test stimuli are highlighted as a representative selection of speech events. Among other questions, the representativeness of speech test materials is analyzed, whereby prototype-oriented and feature-oriented tests in particular are compared. Through studies carried out for this particular purpose, test persons assess speech as a measuring object. Natural as well as synthetic speech sounds are used.

It should be reiterated at this point that it is not a matter of judging the quality of the respective speech synthesis systems. The goal is to systematize, to show which quality aspects the various measurements refer to, to name which criteria they do not consider, and to demonstrate with which test results they each correlate each time [128]. It deals with a functional consideration of speech quality test procedures, with the systematic consideration of factual correlations, whereby neither an assessment of the individual tests is made nor are any recommendations given. This is true also for the so-called CLID Test developed by the author. The CLID Test is presented in greater detail in the discussion than all other tests, but this is due to the fact that the CLID Test is as yet the only test to measure comprehensibility in a feature-oriented procedure. To understand this approach generally more detailed information is needed than to understand prototype-oriented procedures as, e.g. the rhyme test.

10.1 Intelligibility and comprehensibility tests: An introduction

Studies for measuring speech intelligibility have played an important role in speech technology research and development in the past. The question of whether or not a 'technified' speech sound is intelligible depends on many factors. Intelligibility is determined by quality elements of the source, the transmission channel and the information sink. Many users of test procedures in the field of speech technology have as their goal the exact determination of the characteristics of the source and the transmission channel, which influence speech intelligibility (e.g. in the field of speech synthesis). Data from corresponding tests should indicate which parameters or parameter settings of the system elements must be changed in order to target improvements in the overall system.

Test persons (TPs) serve as the measuring apparatus in intelligibility tests. Depending on the test object, various speech sounds are transmitted over different channels (for example, via headphones) to the test persons. The task of the TPs is to describe or assess what they hear. Various procedures are introduced in the following chapters. The essential questions are:

• What intelligibility tests are available?
• What do they intend and what do they, in fact, measure?

10.1.1 The Rhyme Test by Sotscheck

The Rhyme Test reported on here [236] is a monosyllable test for the German language (rhyme tests for other languages cf. page 98) which tests the intelligibility of semantic bearing words, mainly of the phoneme structure consonant-vowel-consonant (or CVC). The vocabulary is phonemically balanced. Therefore, the test can serve as a prototype-oriented test. It is a closed test: The TP is offered six possible answers visually (in a so-called ensemble), whereby one corresponds to the acoustic stimulus given. The intelligibility of either the initial, medial or final word segment is tested per stimulus. For example:

Schuss	Bus	Guss	Kuss	Muss	Nuss
bis	Bass	Bus	büß	bös	beiß
Bach	Bann	bang	Bank	Ball	bald

The Rhyme Test comprises a total of 900 ensembles, 300 each for initial, medial and final word segments. The base material is 600 words.

Assumptions on the meaningfulness of the test results:

• The TPs know in advance that they will hear semantic bearing monosyllables. This knowledge involuntarily guides speech perception (anticipation). The TPs can reconstruct difficult to identify or even unidentifiable segments in the speech signal flow by their combinatory competence. Acoustic elements may not be identifiable as specific speech sounds, nevertheless, the word as a whole may perfectly be intelligible. This is supported by the following results.

• Pilot studies have shown that the same TPs have identified an acoustic signal as a semantic bearing word with the Rhyme Test, while identifying the same signal as a non-semantic bearing word in the context of an open test which consists of both semantic bearing and non-semantic bearing stimuli. That indicates that the method of testing determines, to a considerable degree, the description of what is heard.

• If one intends to make analytical statements about the possibility to identify acoustic signals as speech, it is desirable, in some cases, that listeners give a description of their auditory events unadulterated through a cognitive process. The goal is to dependably identify which signals lead to which perceptive object without it being interpreted to a significant degree by knowledge of the situation and language.

• Not only speech perception, but also the encoding of the auditory events is influenced by the method of testing in the Rhyme Test. When TPs hear a non-semantic bearing word, they are forced, by the test design, to couple it with the most phonetically similar example in the visually offered ensemble. As interviews with TPs have shown, this occasionally occurs consciously due to the nature of the forced multiple-choice test.

• The Rhyme Test is an assimilation-strengthening test. The input signal is in part »translated« rather than »transformed« or »converted« (cf. page 73): The speech signal presented is internally reduced and translated until it fits a visually offered sample. From this follows that diagnostic data which TPs could deliver are lost. In addition, to make matters more difficult, many TPs are conscious of this and become insecure of behaving as expected.

10.1.2 The »SAM Segmental Test«

The SAM segmental test [111] is a comprehensibility test that measures the identification of single consonants. It is a prototype-oriented test. It exists for the German, English, French, Italian, Dutch and Swedish languages. The test vocabulary contains all single consonants which, per language, can appear in word initial, medial *and* final positions, each combined with three vowels (|a|, |I|, |U|) to CV, CVC, and VC structures. Examples follow from the test material:

|ata|, |pI|, |Uk|, |pIp|, |fIf|, |mam|, |an|, ...

Assumptions on the meaningfulness of the test results:

• The test is aimed at the evaluation of the identification of very few consonants. Thus only an extremely limited fraction of all sub-elements possibly generated by synthetic speech are dealt with.

• It remains questionable which usage or application standpoints are supported by the test results concerning these specific sub-elements. Even if the aim of testing should be to compare speech synthesis systems of various languages, an ordinal scale can only be used on the basis of research results if no other conclusions regarding the speech intelligibility of the different speech systems are made. The validity of the test results remains exceedingly limited – if not doubtful.

• It remains to be noted that the individual consonants tested (phones) can only be ascribed the function of allophones of a phoneme intrinsically for each language. Thus, even an ordinal scale of various synthesis systems is not possible since the basis for comparison varies, or can vary, from language to language.

• In testing speech synthesis systems, the problem arises that not every synthesis system has a phoneme input interface. Therefore, in the case of a reading machine that requires a written text as input, it is necessary to transfer the phonemically coded stimulus first into a written representation. Depending on the nature of the internal processing algorithm, it can be difficult, if not impossible, to guide the system to actually produce the desired sound. As a result, the quality of the generated stimulus depends on the quality of the system control.

• The task of the TP is to write down the identified consonants. The difference between voiced and voiceless consonants is not necessarily known to the TPs. The consequence is that cases can occur in which voiced consonants (<tad>) are written down, but which are assigned to the voiceless consonant |t|. This means that the transcription of the TPs must be translated once more. This can lead to misinterpretations or overgeneralizations and thus be additionally erroneous.

10.1.3 The Semantically Unpredictable Sentences SUS Test

The test with Semantically Unpredictable Sentences (SUS) was developed to test word intelligibility in a sentence context. The advantage over single word stimulus tests is the inclusion of prosodic information according to a uniform scheme for all European languages (see [13], [111], [250]). For the preparation of the test vocabulary five sentence structures are set. They serve as patterns for the mechanical generation of sentences. Each sentence is composed of a maximum of six entities (so-called »chunks«) that are filled by a program-controlled, random list access for the respective test.

This test is a morphologically and syntactically prototype-oriented test. Automatic generation has the advantage of disrupting predictability due to semantic information; however, the disadvantage of this approach is that syntactic components can continue to be used by TPs to aid identification. The sentence samples include three declarative constructions, one question and one imperative. The transitivity criterion is taken into account; that is, transitive verbs are only coupled with direct objects, for example. If necessary, so-called mini-syllabic words are formed. A detailed description can be found in [64], [89]. The module that generates the sentences accesses databases with entries selected according to special criteria (for example, according to frequency of occurrence or according to phonetical or phonological factors). Such databases exist for the German, English, French, Italian, Dutch and Swedish languages.

Examples of sentences generated according to the sentence patterns:

English:	*German*:
The hair stands on the light tooth.	Das Haar steht auf dem leichten Zahn.
The proud mouth sees the skin.	Der stolze Mund sieht die Haut.
The door knows the clock which drinks.	Die Tür kennt die Uhr, die trinkt.
Get the train and the stone!	Hole den Zug und den Stein!
How does the chair build the steep wheel?	Wie baut der Stuhl das steile Rad?

Assumptions on the meaningfulness of the test results:

• The intelligibility quota of SUS sentences is not based on the intelligibility of individual words. Individual studies [125] show the following: If a TP does not understand a noun, identifying in its place another one, the tendency is to establish local congruency. It is apparent that in this case semantic and syntactic knowledge of the language significantly influences the response.

• The frequency of occurrence of words used in the sentence test material has an influence on the recognition quota. This becomes a crucial problem if different speech synthesis systems are to be studied and compared to one another. If the same sentence test material is used again and again, memory and learning come into effect. To avoid this, it is desirable to have varied sentence test material available. Results can be compared on the condition that these demonstrate the same degree of complexity: That is, the same degree of analytic listening and identification are needed. Therefore, it is not sufficient simply to expand the entries in the lists accessed by the sentence generator.

• Results show that phonetic aspects are apparently treated with a higher priority than the syntax. The structure is aligned with the sound, i.e. assimilation takes place. Only in rare cases phonetics is secondary to the syntax (accommodation).

10.1.4 The CLID Test

With the Cluster Identification Test (CLID) the comprehensibility of monosyllables is tested [127]. The monosyllables consist mainly of meaningless sequences of sounds of the structure $C_{pre}VC_{post}$, with C_{pre} (pre-vocalic sequences of consonants) and C_{post} (post vocalic sequences of consonants) representing single consonants as well as consonants strung together (so-called consonant clusters). The vocabulary for this test is not necessarily defined. There is a reference list of items (CLID_total), but test items can be generated anew, following parameters set by the person who wants to run the experiment. Thus the form as well and the function of the test changes according to the parameters set. The CLID Test is a feature-oriented test to determinate segmental comprehensibility of consonants. It is an open response test. One word from the test vocabulary is given acoustically and the TP must transcribe what is heard according to the following instruction:

> »Write down what you hear in such a way that, if someone reads it back to you again, he or she would convert your notations into spoken language in such a way that you would identify the same sequence of sounds.«

Thereby, the TPs have two modes of answering available: Either they enter the speech in a phonemic transcription or they describe it in normal spelling using the letters of the alphabet. Per stimulus, syllable comprehensibility as well as the comprehensibility of the pre- and postvocalic clusters and the nucleus, i.e. the vowel, are judged. This, however, in a limited manner:

The CLID Test is a consonant cluster test. The evaluation of the ability to identify vowels is only of limited usefulness. TPs have difficulty noting the length of vowels in such a way that it corresponds to their auditory events. In this case German orthography is an impediment: For example, the <i> in <mir> |mi:6| and <mich> |mIC| is a homographic heterophone that is frequently used. Test instructions and sample notations have no lasting effect on breaking this habit. For this reason so-called »cover symbols« are introduced for the CLID Test, focusing solely on the quality of the vowels. Therefore, evaluating vowel identification is exceedingly limited, if in fact not the actual target of this test.

Assumptions on the meaningfulness of the test results:

• TPs process meaningless words differently as meaningful words. That means that intelligibility (which is, in fact, a function of the comprehensibility of utterances and the understanding that recipients bring to these utterances) cannot be judged under the simultaneous presentation of meaningful and meaningless words. Through computer generation, mostly meaningless words are formed, but some could in fact be meaningful words.

• The previous point of discussion leads easily to the next: If TPs are presented with a meaningful stimulus, they process this in a specific way; then they listen to a meaningless word, where it is to be expected that the ability to combine is still effective. A meaningful word is anticipated and TPs tend to allow themselves to be lead not to note what they have heard, but what they believe they have heard. Translations can take place unchecked.

• It is assumed that, in addition to the frequency of the cluster, the syllable structure, or rather the type of word the syllable structure indicates, also has an influence on the identification of entities. Meaningless monosyllables that indicate the 2nd person singular present, the basic form of which already contain a complex consonant cluster (e.g. |Sampfst|), are more difficult to identify as monosyllables showing the familiar structures of nouns (e.g. |vUl|).

Before dealing with result comparisons, the CLID Test will be used as an example to illustrate which different dependences exist between test vocabulary, test procedures and scaling. This requires, however, a detailed description of the CLID architecture.

10.1.5 General information on the CLID methodology

The most important feature of the CLID Test is that few decisions regarding test parameters and their values can be made in the preparatory phase of the measurement. The CLID Test can be described as an editor, with its structure and process regulated, but with parameters that, in part, must be defined to target certain language specific aspects of the measuring object speech. Such parameters are, for example, the composition of the test vocabulary, the presentation of the stimuli (e.g. timed or untimed stimulus interval) or the form of the answer. An analysis of the conditions (functional aspect of the measuring procedure) precedes each administration of a test, determining the individual aspects of the measuring target. On the basis of this description of the target features, it is decided with which means this target can be achieved. In the case of tests dealing with speech, part of this is the selection of the test vocabulary.

The vocabulary used in the CLID Test is only specified in that it is limited to monosyllables. The number of stimuli is open as is the form of the individual words themselves. The test vocabulary is computer-generated specifically for each application, based on the input from the person who wants to run the experiment and according to the feature description of the measuring target. For example, if system developers know that pre-nuclear plosives are hard to identify in their system, the study leader can generate the test vocabulary in such a way that only pre-nuclear plosives (imbedded in monosyllables) are tested. In this way, suitable data are collected, on the basis of which systems can be improved in specific areas. In another case it could be desirable to set the test material in such a way that the study supervisor gets a differentiated picture of which different micro-characteristics of the speech signal the listener can recognize.

Thus, in principle, the vocabulary of the CLID Test is not predefined. Depending on the target of the study, it can be generated by the study supervisor based on individual features. The vocabulary is generated by a so-called »word generator«. The syllable structure and/or its frequency of occurrence is selected (for example, C_3VC_1 as in the word <Strahl>) as well as consonant cluster structures and/or their frequency of occurrence (e.g. |pfRa:s|), both under consideration of, among other things, statistical information of language and speech. The study supervisor can also define the material independently without using this module (for example, if the intelligibility of meaningful words is to be tested).

In the development of the CLID Test, experiments included a fundamental look at applying the CLID Test procedure to English and French languages. For this purpose, corresponding monosyllable lists were put together. Studies have shown that the CLID approach is, in principle, applicable to French and English languages. There is also an application of the methodology for Korean language [137]. A better representation of the syllables generated would certainly be achieved if the transition matrices included not only yes/no information (cf. Table 10.3), but also frequency data for CVC-sequences. This is true for the German language as well.

The test vocabulary or stimulus material is created according to features by a word generator (WORDGEN). The words generated are phonemotactically correct; that is, they sound like words from the target language but do not necessarily have meaning. The speech sounds are typical for the target language. A non-native speaker learning that language would most likely classify them as words the meaning of which is unfamiliar [10], [134]. In the following, examples for the German language will be given.

WORDGEN is a system that essentially uses three types of data: syllable structure and their frequencies (cf. Table 10.1), clusters and their frequencies (cf. Table 10.2) and matrixes containing information about the syntactic rules of pre-nuclear clusters with nuclei (Table 10.3) or nuclei and post nuclear clusters. For the German language there are 7 basic structures which form combinations to produce all syllable structure types, namely VC_1, VC_2, VC_3, VC_4, C_1V, C_2V, and C_3V. For the generation of the test vocabulary the following parameters are set:

Table 10.1 Syllable structures and their frequency of occurrence

structure	absolute frequency of occurrence	relative frequency of occurrence
C_1VC_1	1.222	21.522
C_1VC_2	926	16.309
C_2VC_2	805	14.178
C_1VC_3	741	13.050
...	...	...

Table 10.2 Consonant clusters and their frequencies of occurrence

cluster	absolute frequency	relative frequency (as to position)	relative frequency (as to number of all clusters)
b	202	3.561	1.209
bl	74	1.305	0.443
bR	7	1.305	0.443
d	14	2.627	0.892
dR	66	1.164	0.395
fR	57	1.005	0.341
...	...	...	...

Table 10.3 Syntactic rules for C_3V syllables

	SpR	Spl	StR	pfR	pfl	skR	skl
a:	x		x				x
u:					x		
e:	x		x		x		
2:			x				
y:							
i:				x			
o:		x					
E:	x						
a	x		x		x		x
...	...	...	...	...	...	...	...

- the extent of the vocabulary (number of words)
- the syllable structure, either directly (e.g. selection of the structure C_3VC_1) or indirectly through the determination of the range of frequency in which the structure(s) should belong
- the frequency of occurrence of the syllable structure of set clusters, divided according to pre- and postinaugural clusters and nuclei

The generated test items are printed out in phonemic notation. This procedure allows test vocabulary to be quantitatively and qualitatively verified. The information describing the test material can be used to purposefully generate test items with certain characteristics. Either test words can be selected in such a way that, in their acoustic form, they carry very discrete phonetical differences or they can be created in such a way that, through their acoustic form, only gross feature differentiations are paid attention to. This test characteristics is extremely important in audiology when different hearing impairments are to be dealt with. The automatic generation of words is attached to a selection module. This module carries out a targeted selection of features through which the vocabulary should be characterized. At this point possibilities to modify the program are available once more. Consequently, at this stage the program supplies automatically produced, mostly non-semantic bearing words, phonemically coded.

After the first processing step the word generator has created the phonemically coded test vocabulary, the phonemically coded words must each be converted into a »graphemic sequence«. The graphemically transcribed form is the form that is used, in general, for reading aloud, i.e. to create the acoustic speech signal. This applies to the people who read the vocabulary aloud in the test preparation phase as well as to text-to-speech synthesis systems. In the case of speech synthesis with phoneme input, test words can certainly be sent directly to the system in the phonemically coded form. In spite of this, however, the conversion step must take place for synthesis systems with a phoneme input interface as well because – as will be shown later – this step is necessary for the automatic evaluation of the results given by the TPs.

As a well-known fact, in the case of semantic bearing words, conventions stipulate how the acoustic-auditory form of individual words are transcribed, either phonemically or graphemically. These conventions are systematically filed as a reference databank for WORDGEN. One use of this reference is to convert the automatically generated phoneme words into grapheme words. Consequently, the program supplies the following material:

- phonemically coded word, mostly non-semantic bearing
- graphemically coded word, mostly non-semantic bearing

After the test vocabulary exists on the symbolic level, the corresponding acoustic signals are produced. The decisions as to which source should be used, through which channel the signals should be sent and which listeners should be taken as TPs are dependent on the purpose for carrying out the test. The following cases serve as examples:

target: testing of a speech synthesis system
source: speech synthesizer
channel: headphones
receiver: test person with normal hearing

target: testing a speech coding device
source: natural speech
channel: speech coder
receiver: test person with normal hearing

target: testing speech identification abilities of patients
source: natural speech
channel: headphones
receiver: listener with speech identification loss

There are various possibilities to administer the actual test. Essential criteria include:

- the selection of the test persons
- the manner of presenting the speech signals (defined level yes/no, fading signal yes/no, defined stimulus interval yes/no, ...)
- the layout of the possible answers for the TPs (correction possible yes/no, input of the »correct« answer(s) yes/no,...).

The decision of how to administer a test in detail is in turn dependent on the test object. Therefore, test conditions are set depending on the test object. Every parameter set-up is recorded by a test program.

The test results are automatically evaluated [170]. Various types of data are available for the system:

- automatically generated words, mostly meaningless, phonemically coded
- automatically generated words, mostly meaningless, as sequences of letters
- per stimulus word one answer (also no response given), coded either phonemically or as sequences of letters

For a better understanding of the algorithmic description of the data selection module, the following example is discussed:

speech signal transmitted	/tra:s/
answer of TP	<drahf>
reference database:	
phonemically coded test word	|tra:s|
with cluster border markings	|-tr-a:-s-|
graphemically transcribed test word (system input)	<trahs>

At this point it is checked whether the answer given by the test person is classified as correct. Since there is no convention for transcribing non-semantic bearing words, there is available a database which lists allowed grapheme sequences for each consonant cluster, depending on its position in a syllable. For example, the syllable initial |tr| can be represented as <tr> or <thr>, the vowel |a:| can be transcribed as <a>, <ah> or <aa> and finally the |s| in the syllable final position as <s>, <ss> or <ß>. Applied to the example above, this means that the pre- as well as post-nuclear clusters in the given answer <drahf> are false. This information is stored in a separate file. If it is useful in terms of the measuring target, the assessed answer from the test person can be shown on the screen directly after inputting the transcription (this can be useful, for example, if the effect of learning or adaptation is to be studied). After processing all stimuli, a so-called response database is created where falsely and correctly generated (or recognized) items are marked. By using this statistics program, different items can be evaluated from various standpoints.

Up to this point, the CLID paradigm has been described. The following now compares different tests for comprehensibility and intelligibility with one another in order to work out the status of each approach. Here the main emphasis is on viewing test results in relative terms to the procedures presented previously in this chapter: the Sotscheck Rhyme Test, the SAM Segmental Test and the SUS Test. For this purpose, various studies for cluster identification and intelligibility of natural and synthetic speech will be introduced. The focus is studying the influences of vocabulary on results. Amongst other things, the CLID vocabularies have been created according to specific language criteria.

10.1.6 Test vocabulary as a measuring influence

All CLID vocabularies cited in this chapter contain 600 entries (called CLID_total). With various syllable structures, typical of certain word classes, the influence of vocabulary on identification results of consonant clusters is studied. In addition, CLID Tests were carried out with monosyllables the structure and cluster frequency of which are typical for all

- entries from the monosyllable list (without words in dialect, technical expressions, first names and place names) (CLID_total)
- nouns (CLID_nouns)
- verbs (CLID_verbs)

Table 10.4 Identification quotas of consonant clusters as a function of word class

	CLID_total			CLID_nouns			CLID_verbs		
	C_{pre}	C_{post}	C_{tot}	C_{pre}	C_{post}	C_{tot}	C_{pre}	C_{post}	C_{tot}
natural voice	91%	86%	**79%**	91%	88%	**81%**	97%	85%	**84%**
synthesis A	88%	67%	**61%**	86%	71%	**64%**	87%	69%	**61%**
synthesis B	79%	75%	**60%**						
synthesis C	80%	71%	**58%**	78%	75%	**59%**	82%	68%	**58%**
synthesis D	88%	68%	**61%**	89%	79%	**71%**	88%	70%	**62%**

All test material was read aloud by a professional speaker as well as generated acoustically with four synthesis systems. Since the speech synthesizers are only the means to an end in this case, the four systems are referred to in the neutral manner of A, B, C and D. Not every voice listed was used to study each individual aspect. The respective tables give further information on this subject.

Each study was carried out with 15 TPs. The sound pressure level (SPL) was adjusted to invariable. The stimuli were presented through equalized headphones. The stimulus interval was not set – as soon as the TP gave an answer, the next stimulus followed acoustically after a short time delay. Using the CLID Test, the TPs were not encouraged to identify a response in every case: If they could not identify a stimulus, they could select a so-called »no response«. Individual TPs took part in two tests at most. The order of the individual stimuli was different per TP for all tests, to exclude adaptation and learning effects in the results. The test vocabulary is differentiated between pre-nuclear consonant clusters (C_{pre}), post-nuclear consonant clusters (C_{post}) and all consonants in the syllable (C_{tot}). The results, given in percentage correct, are listed in Table 10.5. The study results reveal the following trends:

• For all voices and test vocabularies, the pre-nuclear part is identified best and the post-nuclear worst.

• Referring to all consonants within a syllable, CLID_nouns tend to be identified better than CLID_total and CLID_verbs. The exception remains the natural voice: Here CLID_verbs rank before CLID_nouns and CLID_total. (To avoid misunderstandings, it should once again be noted that CLID_total is not comprised directly from CLID_nouns and CLID_verbs, rather that statistic information about the structure of nouns and verbs is included in CLID_total. CLID_nouns and CLID_verbs are not a subset of CLID_total).

- If the differences in the identification results of consonant clusters in dependence of words classified by syllable structure are considered, it clearly shows that there are quite significant differences between CLID_total, CLID_nouns and CLID_verbs (cf. Table 10.5). This is particularly the case for synthesis D, but, in the case of the natural speaker, there are also clear differences regarding the ability to identify »nouns« and »verbs«. It is apparent that the pre-nuclear part of verbs is significantly better identified than all other clusters.

- A similar tendency is seen in synthesis C: Here also pre-nuclear consonant clusters of verbs are identified better in comparison to nouns. It is apparent once again that post-nuclear parts of verbs are identified conspicuously worse than those of nouns.

Based on these data, it can be concluded that consonant clusters of »nouns« in comparison to consonant clusters of »verbs« are identified differently. Observing in addition the frequency distribution of pre-nuclear and post-nuclear consonant clusters, a significant difference is visible. This difference is most likely the cause of the identification differences between CLID_nouns and CLID_verbs.

If one analyzes the CLID Test results further regarding reliability, results as listed in Table 10.6 have to be discussed. There are a few test persons who identify and/or describe what they have identified different than the average of all TPs do. To find an explanation for these results a look in the raw data list is very informative. In the case of TP11, synthesis D (cf. Table 10.7), for example, it is noticeable that pre-nuclear consonant clusters were identified differently as intended by the synthesis. As an example, the regularity with which the initial /R/ is confused with other sounds shows that identification difficulties can clearly be traced back not to unwanted listener characteristics, but to specific characteristics of the acoustic signal which these listeners are able to hear. TP11 transcribed the initial /R/ as <br> with few exceptions. Assuming that there is no error in the transcription method, it can be concluded that she apparently truly had the corresponding auditory event /bR/. The conclusion reasonable to assume is that the auditory form of this synthesized item obviously lies in the space around the dividing line of speech perception between /R/ and /bR/ for TP11 – and this is scaled by the test. In the context of judging quality with the CLID Test, it follows that this result in the end speaks for, rather than against the test. If one qualifies the results and evaluates the data without the cited »rogue result«, minimal, negligible deviations result. Consequently, one may not therefore speak of »identification errors« or »erroneous recognition«, but rather refer to speech signal characteristics that do not lead to the desired auditory perception. (In spite of this, in the following the concepts of »falsely« and »correctly« identified are used for reasons of simplicity – nevertheless, they should be understood as described above.)

Table 10.5 Differences in identification results of consonant clusters in dependence of words classified by syllable structure and cluster frequencies

	CLID_total			CLID_nouns			CLID_verbs		
	C_{pre}	C_{post}	C_{tot}	C_{pre}	C_{post}	C_{tot}	C_{pre}	C_{post}	C_{tot}
natural voice	0	-2	-2	**-6**	1	**-5**	**-6**	3	-3
synthesis A	2	-4	-3	1	-2	0	-1	2	3
synthesis C	2	-4	-1	-2	-3	0	-4	7	-1
synthesis D	-1	**-11**	**-10**	0	-2	-1	1	**9**	**9**

Table 10.6 List of outliers

	CLID_total		CLID_nouns		CLID_verbs	
	C_{pre}	C_{post}	C_{pre}	C_{post}	C_{pre}	C_{post}
natural voice						
synthesis A	TP05: -10%	TP05: -10%		TP17: +21% TP12: -11%		TP15: -17%
synthesis B		TP08: +12%				
synthesis C		TP10: -10%		TP08: -9%		TP04: -16%
synthesis D	TP11: -12%	TP02: -14%	TP11: -10%	TP11: -10%		

Table 10.7 TP11–Extract of results: synthesis D, CLID_total, alphabetic order

test stimulus	item identified		pre-nucleus: sent–identified		nucleus: sent–identifed		post-nucleus: sent–identified	
R-aI-kst	bR-aI-kst	0	R-bR	0	aI-aI	1	kst-kst	1
R-aI-kt	bR-aI-kt	0	R-bR	0	aI-aI	1	kt-kt	1
R-E-m	bR-E-m	0	R-bR	0	E-e	1	m-m	1
R-i:-6	bR-i:-6	0	R-bR	0	i:-i:	1	6-6	1
R-I-k	bR-I-Nk	0	R-bR	0	I-I	1	k-Nk	0
R-I-kst	bR-I-kst	0	R-bR	0	I-I	1	kst-kst	1
R-I-kt	bR-I-kt	0	R-bR	0	I-I	1	kt-kt	1
R-I-ls	bR-I-lst	0	R-bR	0	I-I	1	ls-lst	0
R-I-lt	bR-I-lt	0	R-bR	0	I-I	1	lt-lt	1
R-I-m	bR-I-m	0	R-bR	0	I-I	1	m-m	1
R-O-mst	R-O-mst	1	R-R	1	O-O	1	mst-mst	1
R-U-xt	bR-U-xt	0	R-bR	0	U-U	1	xt-xt	1
R-U-p	R-U-p	1	R-R	1	U-U	1	p-p	1
R-aI-k	bR-aI-k	0	R-bR	0	aI-aI	1	k-k	1
R-aI-kst	bR-aI-kst	0	R-bR	0	aI-aI	1	kst-kst	1

Table 10.8 Examples of insertions at word initial position

CLID_total, synthesis A	
initial clusters (sent in total)	inserted clusters (frequency of occurrence)
E(45)	h(1)
a(90)	b(2), p(2), R(2), h(1), k(1)
aI(30)	b(3), R(2)
i(45)	j(1), l(4)
o(15)	b(5), v(1)
u(30)	R(2), b(6), bR(10), gR(1), h(1), v(3)
y(15)	v(4), z(1)

It must be stressed accordingly that there *can* be noticeable differences between CLID_total, CLID_nouns and CLID_verbs. Study results show that this effect does not necessarily appear for every speech object studied, but *that* it can result and *that* it therefore must be controlled and checked.

The CLID Test is an open comprehensibility test. Up to now the focus of the data analysis was the discussion of confusion of (intended) generated acoustic form and the identified auditory form. The size of reference is the unit »cluster«. The decision whether or not something is identified as »correct« or »false«, is not only based on confusion, but also on additions and omissions. An example:

The word <brasch> is generated, <aparasch> is identified (insertion).
The word <brasch> is generated, <_asch> is identified (omission).

Focusing the data analysis on the characteristics »insertion«, synthesis A stands out in particular. Here mainly initial vowels seem to be characterized by perception features that are classified not as a glottal stop, but as a component of a consonant (cf. Table 10.8).

An additional aspect that is connected to the vocabulary is the dichotomy »semantic bearing« versus »non-semantic bearing«. In the first phase of development of the CLID Test, the vocabulary comprised semantic bearing as well as non-semantic bearing monosyllables. Pilot studies showed, however, that TPs apparently fall more easily into a cognitive mode when semantic bearing and non-semantic bearing words are presented. This is a disruptive effect in presenting non-semantic bearing stimuli since comprehensibility tests are meant first of all to examine whether the speech signal carries sufficient elements to support an unambiguous phone classification. This requires analytical listening. However, in simultaneously presenting semantic bearing and non-semantic bearing stimuli, TPs tended in part to relate to the general auditory form (or to the Gestalt) and to process or respond less analytically

Table 10.9 Extract of results of association experiment

stimulus	TP1	TP2	TP3	TP4	TP5	TP6
blax		blass	Blag	lach	Bach	
blIRS		Hirsch		Hirsch		
bIlt	Blitz	Blitz		glitt	glitt	
bOft				oft		
bOpfs		Kopfs				

in their perceptions of auditory sequences. In this way, they compare the form of the speech perceived – how they habitually process speech – with known semantic bearing speech patterns and identify accordingly semantic bearing words more frequently. For the quality of the evaluation, it follows that it can no longer be assured that in fact comprehensibility alone is measured. On the contrary, it is assumed that speech intelligibility, which is in fact a function of comprehensibility and comprehension, is judged.

This is not a desired effect – or rather: This is an effect that should be examined (because in some research contexts the goal of the study may well focus on this aspect). In order to be able to check the feature »semantic bearing«/»non-semantic bearing« a stimulus list with exclusively non-semantic bearing words was generated with the CLID paradigm as well.

But even when only non-semantic bearing words make up the stimulus material (as is the case in CLID_total), it still has to be proven whether the answers really correspond to comprehensibility. An additional pilot study was instituted to research this aspect: Non-semantic bearing CLID stimuli (natural voice) were presented acoustically to TPs. The task, however, was to associate a semantic bearing word that was most similar to the perceived auditory form of the stimulus. Several examples are listed in Table 10.9. An analysis of the results reveals that isolated answers by TPs in the CLID Test mentioned above are apparently to be explained by a comparable process of association. These, however, are exceptions. Out of the whole of 600 stimuli to be identified, they appear comparatively rarely.

10.2 On the comparability of study results

The form and content of procedures for judging speech quality are dependent on the context in which they are used. There is no »the test« that can be used for all speech events. Furthermore, there are also no universal guidelines that clearly stipulate the configuration of speech quality tests. These and other factors make the universal use of speech quality tests impossible. The unreserved use of tests and their results is further aggravated by a systematic examination of the fundamentals.

The studies of speech comprehensibility and intelligibility referred to in the previous chapter were carried out to profile the correlation of speech utterances and segmental speech comprehensibility. As shown, word class typical phonotactic features of the vocabulary can indeed influence the identification score in one and the same test. Of the voices studied, this effect arose particularly in synthesis D. This is of note because this synthetic voice is always produced according to the same rules, independently of word class typical syllable structures. The fact that the identification score for stimuli of noun and verb structures is also not the same in the case of natural speakers makes clear that, as known, spoken language is in no way a simple stringing together of single sounds. In speech production as well as in speech reception, the form of the perceived individual phones results by far from a more extensive overall context. This leads to truly different acoustic-auditory forms of speech. Alone the mutability of speech forms in various expression contexts presents speech quality experts with an almost unsolvable question, namely:

> Is there a representative subset of speech stimuli that should be included in the test vocabulary, whereby the stimuli are created in such a way that allows conclusions from special sets (i.e. from test results) to apply to the general (e.g. to more extensive vocabulary)? If yes, how many stimuli must this set include and of what type should they be?

At present it is not possible to give a satisfactory answer to this question. If it could be found, not only the field of speech quality assessment, but also research and development tasks for speech technology products could make considerable use of it. Due to the complexity of the aspects posed by this question, it is therefore appropriate to concentrate first of all on *one* specific standpoint and study this in greater detail. This follows:

A series of assessments to determine speech intelligibility is described, the sole target of which is to analyze the interdependence of assessment test methods, vocabulary and scaling in the context of monosyllables. The question of the comparability of results on comprehensibility and intelligibility of natural and synthetic speech is studied. Besides natural speech with up to 5 various Signal to Noise Ratios (SNRs), the CLID Test, the Rhyme Test, the SAM Segmental Test (for reasons of simplicity also referred to as the VCV test), and the SUS Test are used, varying the vocabulary and the response mode.

The following is valid for this series of investigations: All test material was read aloud by a professional speaker as well as acoustically generated with four synthesis systems. The four synthesis systems are designated A, B, C and D. Not every voice named was used in studying every individual aspect. The respective tables provide further information. Measuring objects are the influence:

- of the vocabulary (syllable structure and frequency, semantic bearing versus non-semantic bearing)
- of the presentation as a single word or in the context of a sentence
- of the response mode (open versus closed test)

The enumerated tests and their respective functions are compared schematically in Fig. 10.1. The convention for the choice of names is as follows: The first part of the word names the paradigm (in the present example CLID), the second the vocabulary (vocabulary of the Rhyme Test). In the following, several such combined procedures are described. Several tables of results are included to serve as a basis for the discussion to follow, namely the identification quotas and scores for C_{total} (only pre- and post-nuclear consonant clusters) and the discrepancies of identification results of various tests procedures in Table 10.10.

10.2.1 On the influence of scaling

An experiment is carried out to study the influence of scaling on the recognition score. The Rhyme Test paradigm by Sotscheck is used. In addition, the comprehensibility of all Rhyme Test words (which are semantic bearing monosyllables) are studied with the CLID Test paradigm. According to the convention described above, this experiment is referred to as CLID_rhyme.

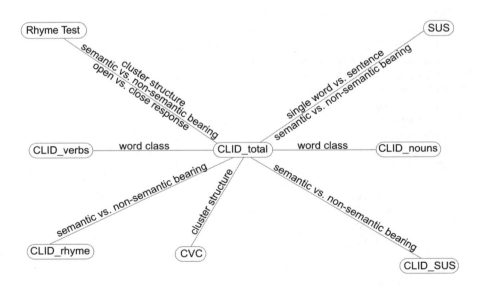

Fig. 10.1. Distinctive features of auditory methods

The study was carried out for natural voice without random noise. The results are as follows: Rhyme Test 100% vs. CLID_rhyme 91%. At first glance, this result is surprising. It was expected that both tests would lead to almost the same results, but the experiment shows that natural monosyllabic words listened in isolation are in fact not 100% comprehensible. An analysis of the test persons' answers carried out by the study leader after the experiment shows that they were explicable or comprehensible through specific features of the speech sounds. These can lead to crossing phoneme boundaries. As the example shows, this is true for semantic bearing words as well. The difference of 9% points between the Rhyme Test and CLID paradigm with Rhyme Test stimuli (CLID_rhyme) clearly shows that the Rhyme Test in the

Table 10.10 Studies in comparison: discrepancies of identification quotas

	natural voice	natural voice, -10dB	natural voice, -15dB	synthesis A	synthesis B	synthesis C	synthesis D
Rhyme Test vs. CLID_rhyme	9						
Rhyme Test vs. CLID_total	21				37	40	38
Rhyme Test vs. CLID_nouns	19					39	28
Rhyme Test vs. CLID_verbs	16					40	37
CLID_rhyme vs. CLID_total	12	13	6				
CLID_rhyme vs. CLID_nouns	10						
CLID_rhyme vs. CLID_verbs	7						
CLID_total vs. CLID_nouns	2			-3		-1	-10
CLID_total vs. CLID_verbs	5			0		0	-1
CLID_total vs. SUS				-8			
CLID_total vs. CLID_SUS				-14			
CLID_total vs. CVC				-25			
CLID_SUS vs. CLID_nouns				11			
CLID_SUS vs. CLID_verbs				14			
CLID_SUS vs. CVC				-11			
CLID_nouns vs. CLID_verbs	-3			-3		-1	9
CLID_nouns vs. SUS				-5			
CLID_verbs vs. SUS				-8			
CLID_SUS vs. SUS				6			
SUS vs. CVC				-17			
CVC vs. CLID_nouns				22			
CVC vs. CLID_verbs				25			

case of this stimulus characteristic is already in the saturation space, i.e. it is no longer meaningful. In contrast, the CLID paradigm aids, even for natural speech without noise, in the identification of exactly such speech sounds, for which no clear phoneme classification is possible. Apparently the Rhyme Test method significantly guides speech perception and identification. Data lead to the presumption that ambiguities and indistinctness of the speech stimulus are solved solely by presenting possible answers visually. Anyhow, the characteristic of the Rhyme Test discussed here must in no way be seen as a disadvantage: The Rhyme Test is an intelligibility test which depends in particular on bringing comprehensibility and comprehension into accord.

10.2.2 On the influence of the stimulus context

This series of studies examines whether the presentation of monosyllables as individual words or in the context of a sentence is to be considered as a dimension of measuring influence. For this purpose the SUS Test is carried out. TPs listen to semantically unpredictable sentences in an open response mode. The task is to write down what is heard. Each sentence is made up of semantic bearing words. With few exceptions, these are monosyllables.

The comprehensibility score of those monosyllables that belong to the SUS vocabulary are additionally studied with the CLID paradigm (CLID_SUS), thus also in open response mode. The test is carried out with one voice (synthesis A). The result: SUS Test 69% and CLID_SUS 75%. With the same synthesized speech material, the same synthesis and the same scale, a difference of 6% points occurs through the intelligibility measurement of stimuli either in the context of a sentence or as individual words. It is presumed that this can be put down to the stimulus *context* alone.

This conclusion is rash since SUS stimuli (separate sentences) are not to be understood as a simple concatenation of individual words, as is studied with the CLID Test. Hence in sentence processing in speech synthesis system A, a syntax module is activated that, for example, marks content and function words in the respective sentence to be synthesized and classifies corresponding speech sound lengths and assigns fundamental frequency. It follows that the form of the synthesized sentences in SUS context are, by all means, something different as the sum of CLID_SUS stimuli.

The determination of the reasons for this scaling difference must lie in something more profound. If the TP results are analyzed in more detail regarding the measuring entity »word correct«, the assumption is confirmed that the difference between SUS and CLID_SUS cannot only be explained by source-specific entities, but that there is significant influence due to the manner in which the stimuli are presented (and therefore, the manner of scaling). To a not insignificant degree, apparently various listener-specific behavior patterns are in effect that can be outlined as follows:

The presentation of semantic bearing individual words in the context of a sentence initiates, first of all, a receptive mind-set for the TPs that corresponds to the behavior usual in speech processing. In general, the strategy in speech processing is to extract the content of the sign carrier with a minimum of listening time by filtering out the important information cues. In other words, receptive behavior concentrates on what it anticipates to be necessary and is limited to what proves to be of advantage for understanding. The answer to the question of what is of advantage and what not has been developed and set in the lengthy course of language acquisition. This learned behavior form is deeply automated and occurs first of all involuntarily.

However, after the TPs have heard the first words of the sentence in the test laboratory, this receptive mind-set is disturbed. The TPs manner of anticipating based on experience and knowledge is suddenly no longer useful in the context of the study. On the contrary, it is rather obstructive and disruptive.

Anticipation, however, still functions as a filter. Usually it minimizes the overall cost of perception load to the extent that speech comprehension is only just guaranteed in combination with combinatory competence and knowledge of the world and experience. In the case of SUS sentences, this sentence is antagonistic to semantics. That is due to the fact that anticipation is targeted at exactly such entities that are *not* just offered acoustically (since they are semantically abnormal). In other words, anticipation limits exactly those entities (in the sense of anticipation of semantic probabilities for transitions) that are typical for SUS Test material.

In addition, the following aspect makes matters worse. TPs perceive semantic bearing words in the context of a sentence in the SUS Test, by which local congruency is ensured. This mixed form of familiar and unusual speech semantic characteristics in the flow of the sentence can lead to an inconsistent manner of speech reception: While TPs have only just distinguished and described what they have heard analytically, in the next instant speech processing can, through certain moments that trigger associations, change into a strongly anticipation-steered interpretation of the content. In regard to the continuum »conversion-translation« and »comprehensibility-understanding«, a constant shift of focus can take place. Analysis of data materials and interviews with the TPs verifies that such processes and reactions do occur.

Therefore, it can be assumed that the difference of 6% points between SUS and CLID_SUS results can be traced back, to a considerable degree, to inconsistent behavior by TPs. However, since it cannot be fully ruled out that a certain influence of source-typical features is mirrored in the test results as well, no clear determination of the cause can be found.

10.2.3 On the influence of syllable structure and frequency

The focus of the series of studies described here is the question of which conclusions can be drawn from a comparison of intelligibility and comprehensibility measurements. Up to this point, analysis of the results has shown that the method of presen-

Table 10.11 Identification results Rhyme test vs. CLID_total

	Rhyme test	CLID_total
natural voice	100%	79%
synthesis B	97%	60%
synthesis C	98%	58%
synthesis D	99%	61%

tation of stimulus elements of the same content as well as the response mode have a considerable influence on the test results. In view of this dependency, it is to be expected that much greater differences will appear if in addition to this influence the vocabulary is also not identical (but comparable from a structural point of view). In order to study this question in greater detail, results of the Rhyme Test and CLID_total paradigms will be compared. The Rhyme Test and CLID_total differ regarding vocabulary in terms of semantics (semantic bearing/non-semantic bearing) and syllable structure (CVC vs. C_nVC_n) as well as scoring (closed/open response mode). The results (for speech without noise) are as follows:

Natural speech shows a difference of 21% points; synthesis B, 37%; synthesis C, 40%; and synthesis D, 39% points. It stands out that all CLID_total results of the synthesis systems studied lie in the area of 60% with a slight mean variation (cf. Table 10.11). However, CLID_total with natural speech stimuli without noise reaches 79% and sinks to 9% with increasing SNR conditions (cf. Fig. 10.2.). Calculating the relation of the distance (Rhyme Test vs. CLID_total), it lies at 1.3 for natural speech of 0 dB SNR (cf. Fig. 10.2.), for the synthesis systems at 1.6 (B), 1.7 (C) and 1.6 (D) (cf. Fig. 10.3.). These data show two things:

- There is a considerable difference between the Rhyme Test and CLID_total results.

- Regarding the relation of the distance between the Rhyme Test and CLID_total, there is a noticeable difference between the measuring objects »natural speech« and »synthetic voice«.

Regarding the first point: As has been shown the difference between the Rhyme Test and CLID_total can be explained by the influence of the response mode. In addition it is very likely that identification results are influenced by vocabulary. A comparison of the structure of Rhyme Test and CLID_total stimuli makes this clear in regard to the structure of the German monosyllable list:

While the Rhyme Test takes into account the structures C_1VC_2, C_1VC_1, C_2VC_2, C_1VC_3 and C_2VC_1 only, the vocabulary of CLID_total reaches almost full agreement with the structure types of German monosyllables (cf. Fig. 10.5.). The distribution is adjusted to structure occurrence and frequency. The Rhyme Test in contrast uses

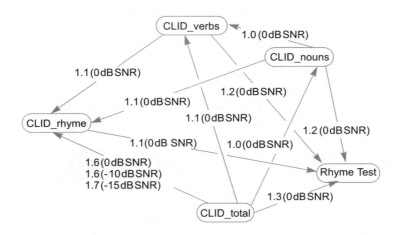

Fig. 10.2. Distance relation: different procedures, natural voice with different SNRs

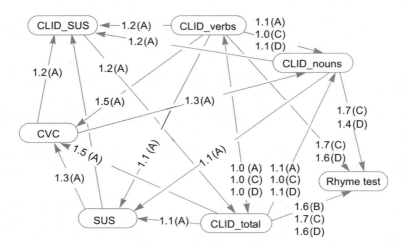

Fig. 10.3. Relation of the distance of different investigative procedures using 4 synthetic voices (A, B, C, D)

mainly stimuli of the structure C_1VC_1. This structure is represented 4.4 times less frequently in the monosyllable list as in the Rhyme Test, while the structure C_2VC_2, is underrepresented in the Rhyme Test in comparison to entries of the monosyllable list. Consequently, the Rhyme Test makes use of stimuli that cover neither the scope of possible syllable structures nor their frequency in regard to the monosyllables of the German language. This is also true for subsets of syllable structures such as the set of German monosyllabic nouns. Further statistical details are presented in Table 10.12.

Apart from the structure, the architecture of a syllable is also determined by its components. Thus various consonant clusters can occupy the pre-nuclear as well as the post-nuclear parts of monosyllables. Fig. 10.6. lists the respective pre-nuclear clusters and their frequencies for German monosyllables for the CLID_total and Rhyme Test vocabulary. At this point a comparison of the pre-nuclear clusters should suffice. The graphics show considerable differences between the two tests regarding the frequency of occurrence of individual clusters, which could also be a reason for the inconsistent results (cf. Fig. 10.6.).

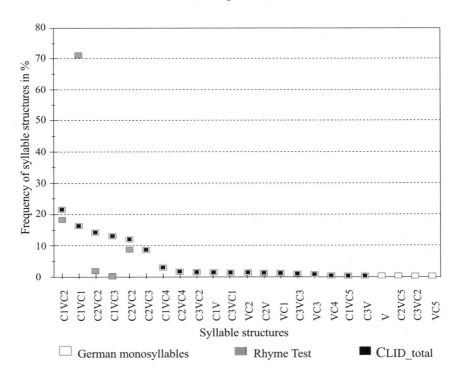

Fig. 10.5. Frequency of syllable structures: German monosyllables, CLID_total and Rhyme Test

Table 10.12 Structure and frequency of occurrence in % of monosyllables in German vocabulary and in lists of test stimuli

syllable structure	semantic-bearing (all entries)	semantic-bearing nouns	CLID_total	CLID_nouns	Rhyme Test
C_1VC_2	21.522	24.092	21.833	24.167	18.216
C_1VC_1	16.309	27.151	16.000	27.333	71.216
C_2VC_2	14.178	10.516	14.167	10.500	1.708
C_1VC_3	13.050	4.685	13.000	4.667	
C_2VC_1	12.117		12.167		8.918
C_2VC_3	8.894	2.103	8.833	2.167	
C_1VC_4	2.888	0.096	2.833		
C_2VC_4	1.656		1.667		
C_3VC_2	1.391	0.956	1.333	1.000	
C_1V	1.338	2.199	1.333	2.000	
C_3VC_1	1.286	2.103	1.333	2.167	
VC_2	1.215	1.912	1.333	2.000	
C_2V	1.074	1.816	1.167	1.833	
VC_1	0.951	1.530		1.500	
C_3VC_3	0.845	0.096			
VC_3	0.652	0.382		0.333	
VC_4	0.123	0.096		0.167	
C_1VC_5	0.123				
C_3V	0.123	0.287		0.333	
V	1.106	0.096			
C_2VC_5	0.088				
C_3VC_4	0.053				
VC_5	0.018				

With regard to the dichotomy semantic bearing /non-semantic bearing, syllable structure occurrence and frequency as well as cluster occurrence and frequency, the Rhyme Test and CLID Test measure quite different aspects of monosyllables.

10.2.4 On the validity of the test results

In the previous chapter, various investigative processes for measuring speech comprehensibility and speech intelligibility were introduced. Discussion of the results shows that different processes also lead to different results. This is the case although all investigative processes are seemingly targeted at measuring the comprehensibility and intelligibility of speech sounds. If the individually measured values are now considered as estimated values for the measurand, the question of the true value of speech comprehensibility or speech intelligibility inevitably arises. In metrology, the concept »true value (of the measurand)« is defined as follows [46]:

DIN Def. »true value«

"value of the measurand [...] as target of analysis of measurements [...] of the measurand."

Precisely the comparison of results from CLID_total, CLID_nouns, and CLID_verbs shows that the existence of a definite value and therefore the true value for that measurand is guaranteed that in fact exists under the conditions prevailing during the measurement: CLID results are reproducible for non-semantic bearing monosyllables without interfering noise according to the differentiation no word functional

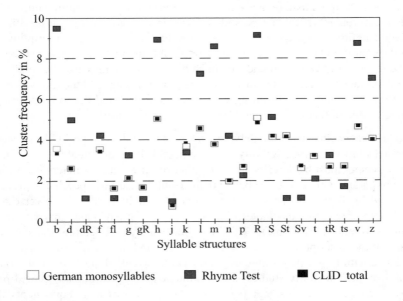

Fig. 10.6. Pre-nuclear consonant clusters of Rhyme Test words and their frequency of occurrence in German monosyllables, in CLID_total and in Rhyme Test vocabulary

difference (CLID_total), noun and verb. In order not to jeopardize the validity of the results of measurements, this must without fail be entered into the description of the CLID measurand as follows:

The measurand for CLID_total is the comprehensibility of non-semantic monosyllables that, according to their structure and cluster frequency, depict the features of all German monosyllables (without inclusion of their respective frequencies of occurrence in language use and without loan and dialectal words, technical expressions, given names and geographical names). The validity of the CLID Test is only proven for this aspect.

A principle of psychometry says: Only if each measurand in each measurement procedure is determined and described unambiguously, it can be guaranteed that each individual procedure can suffice in reliability and validity. Aspects that can be applied to validity and reliability examinations were discussed in detail in the previous chapters. In addition, results of further tests and experiments were analyzed in comparison to the CLID Test. These analyses allow for the assumption that test results of other procedures are also reproducible. The unalterable precondition is, however, that the descriptions of the respective measurands are determined in detail. This necessity is justified in [46] as follows:

> "The existence of a definite value and therefore the true value is guaranteed for that measurand that, in fact, exists under the conditions prevailing during the measurement. If the measurand is determined in a measuring task, it is given a definite value and therefore a true value, in so far as the description of the measurand is complete. If the description is incomplete [...], the value or true value of this measurand cannot be referred to. In this case, the conditions set incompletely in the measuring task are fulfilled, and the true value of the measurand present during measurement is declared the to be determined value of the incompletely defined measurand." [46]

However, exactly that unfortunately still takes place today in the field of speech comprehensibility and intelligibility measurement: It is ignored that the measuring object »speech« must be described by a number of different parameters since each change of parameter can bring a change in the results regarding speech comprehensibility and intelligibility. This chapter presented a study of selected, diverse features solely from monosyllables, based on a minimal pair analysis in a communicatively neutral context. This identified exactly such features that contribute only to the description of the respective measurand in this specific context. In accordance with linguistic terminology, they could be called »testemes« of monosyllable-comprehensibility studies in a communicatively neutralized, guided perceptive context. What must not occur, however, is to treat as equivalents the estimated values of incompletely described measurands and the true value of comprehensibility of a voice x.

Therefore, the results of one single test cannot be conclusive for the comprehensibility of the voice in general. One cannot conclude that synthesis x is better than synthesis y because synthesis x scored higher in a comprehensibility test. However, all test results may well be used to identify the weak points of a particular system and to undertake targeted improvements in the system based on these data.

10.3 Summary

Diverse procedures for measuring comprehensibility and intelligibility were presented and test results were compared. It was shown that:

• The CLID Test is a segmental comprehensibility test for the entity »monosyllables«. The measurands are pre- and post-nuclear consonant clusters. This test stands out from all the others in that the vocabulary is generated based on a defined set features. It is a feature-oriented test. That means that the vocabulary is not set, but rather it can be generated specifically according to features.

• Reproducible results are achieved with the test. It is a precondition, however, that the description of features is identical for the repeated application.

• It was shown that the structure of the monosyllables (whether nouns or verbs) has an influence on the results: Consonant clusters from nouns are identified differently than consonant clusters from verbs. This is true not only for the synthetic, but also for the natural voices.

• The CLID Test paradigm offers a form to scale not only substitutions, but also omissions and insertions (in regard to what was intended to acoustically generate/ auditorily perceive).

• Studies have shown that it is advisable to separate semantic bearing and non-semantic bearing stimuli clearly. TP performance is too inconsistent and not verifiable if both elements in vocabulary are present in one and the same test.

• The aim of the discussion was to analyze several fundamental principles of speech quality assessment. To achieve this, a number of different studies were carried out. Among other things, they refer to an examination of the quality of known procedures for assessing speech comprehensibility and speech intelligibility. This showed that the procedures considered produce each time reproducible results, but that the task of measurement is frequently not adequately specified. Thus *the* speech comprehensibility and *the* speech intelligibility cannot be measured, but only certain aspects thereof. That means that test results must always be qualified: They are functional in regard to the test vocabulary and response mode.

11 The Cluster Similarity Study

Paradigm selection plays an essential role in speech quality measurement. Among other things, specifications of language and speech test material make up part of the respective paradigm. The markedness of different features in the test material influences the test results in a significant way. The studies on speech comprehensibility and intelligibility in the previous chapters document this. They focus mainly on linguistic questions. Essentially, from the point of view of perception they consider each speech element as being equally distinguishable. Anyhow, we know that a /m/ (/maIn/ <mine>) is perceptually much closer to a /n/ (/naIn/ <nine>) than it is to a /S/ (/SaIn/ <shine>). In the intelligibility context, a confusion between <mine> and <nine> is not as severe as a confusion of <mine> and <shine>. This has to be encountered for both defining the speech test material for intelligibility tests as well as analyzing and interpreting test results.

In this chapter as well, linguistic questions on the measuring object »speech« are considered, but from a different standpoint. Up to now, perceived form characteristics of the measuring object with regard to their ability to carry content were to the fore: However, now the aspect of form–content does not play a role any more but focus lies exclusively on the form itself. The study deals with portraying form characteristics of content-neutral speech test material with regard to their perceptual similarity. Linguistic standpoints are certainly taken into consideration, however, solely regarding phonologic and phonotactic criteria of speech test material specifications. Additional characteristics do not play a role.

In this chapter the so-called »cluster similarity study« is described. The approach is to compare – in reference to the context – speech sound sequences. For the preparation of the speech material target syllables are embedded in the same position in carrier sentences, one for pre- and one for post-nuclear consonant clusters (the sentences were set at the Institute for Communication Research and Phonetics in Bonn for the selection of speech components for speech synthesis):

"Das wäre **ta**telei gemacht."   – **ta**   is the target syllable.
"Das wäre **ma**telei gemacht."   – **ma**   is the target syllable.
"Das wäre **stra**telei gemacht."   – **stra** is the target syllable.

...

"Das Stoßgeb**at** ist ohne Sinn."   – **at**   is the target syllable.
"Das Stoßgeb**am** ist ohne Sinn."   – **am**   is the target syllable.
"Das Stoßgeb**anf** ist ohne Sinn."   – **anf** is the target syllable.

...

With regard to the test material, the choice is put on a »prototype-oriented representation« of form-related elements. The aspect of prototype-oriented representation requires a uniform basis for comparison. 'Snap shots' of speech acoustically produced according to identical model patterns are taken and examined regarding their perceived phonetic similarities. The model patterns can be described as follows: The target syllables are always in the same place in a carrier sentence of a typical sentence structure. The carrier sentence is made in such a way that when read aloud there are no essential coarticulation effects. Therefore, giving acoustical form to the target syllable for the production of speech vocabulary takes place through its respective syntagmatic relation in sentence context.

To produce the speech material, each sentence is spoken aloud and recorded digitally. Out of the acoustic speech signals (i.e. the sentences), speech 'slices' are produced. These are the monosyllables already mentioned above (e.g. /ta/, /ma/). As in the studies for speech comprehensibility and intelligibility, natural speech and synthetic speech are used as sources. The 'slices' of monosyllabic speech sounds make up the stimulus material for the cluster similarity studies. Only the perceived similarities of consonant clusters in dependence of specific vowel contexts are assessed.

The results of this approach are only valid under the defined conditions. It is a well-known fact that the form of speech sounds is also characterized by factors beyond the entity »sentence«. Furthermore, factors outside of the realm of linguistics have an effect of course. Simply the fact that the speech excerpts are not spontaneously spoken but read aloud, shapes the form of the speech sound in certain ways. The interpretation of the results must, therefore, occur with care under consideration of all factors influencing measurement and may not generalize uncritically. Within the set limits, this approach does uncover quite impressive tendencies that contribute to a complete understanding of speech quality events.

To sum it up: The aim of the study is to portray perceived similarity of speech elements, namely consonant clusters.

11.1 On the background of the study

The starting point of this study is the two realizations that naturally spoken words are not always 100% comprehensible and that even 100% comprehensible, synthetic speech may only have an extremely slight acceptance. An analysis of comprehension test results of natural speech sounds clearly shows that the possible range of variations of all allophones of a phoneme is, on the one hand, quite variable, but, on the other hand, rather restrictive at the same time. Thus the pre-nuclear phone /d/ in the word <das> is in no way *the* sole acoustic realization of the phoneme |d|, but under certain conditions could belong to voiceless /t/ variants in the allophone inventory of the phoneme |d|. If this voiceless /t/ variant is then perceived auditorily e.g. in the CLID context, they are easily identified as allophones of the phoneme |t|: The reason for this is that there is no further contextual language information that helps to resolve the indistinctness between the allophone sets of |d| and |t|. In other words, in fluent speech, meaningful identifications manifest themselves in phonologic, grammatical and lexical contexts. Without additional information, such as semantics and pragmatics, no distinct assignment of a perceived sound and its classification is possible. There certainly are sets of allophones that can be assigned, in such a neutral context without additional information, to different phonemes. This is not surprising in so far as the classification of allophones of phonemes results from the so-called minimal pair or commutation test. With this test it is studied whether a phonetic difference between two speech sounds embedded in the same context achieves a change in meaning. If the substitution in the same context leads to a change in meaning, both speech sounds are assigned to different classes of sounds, whereby the sounds classes are called »phonemes«. If it leads to no change in meaning, they are classified as »allophones of a phoneme«. A feature of the allowed overlap formation of allophones appears to be the perceived similarity of phones. This overlap is a natural phenomenon. The design of a synthetic voice which is characterized by high naturalness must model this. The goal of this study is, therefore, to measure the similarity of selected phonetic items.

In linguistics one can roughly differentiate two methods to describe aspects of speech: in a diachronic or synchronic approach. The diachronic refers to the development of speech and language in contrast to the synchronic one that concentrates on one point in time. The classification of speech sounds described above has already been quite thoroughly examined in linguistics from the diachronic standpoint. Jakobson, for example, made a significant contribution in identifying and describing regularities in phone and phoneme systems in regard to ontogenesis (child language acquisition), phylogenesis (evolution of speech systems) and pathogenesis (language loss) (cf. [123], [23]). The following presents a detailed look at models of child language acquisition in order to make clear its significance to this theme:

In child language acquisition, the pre-linguistic phase of cooing and babbling is differentiated from the phase of the first speech sounds. In contrast to cooing and babbling, speech sounds get a sign function. While the child uses speech organs spontaneously in the pre-linguistic phase and produces enormously wide variations of sounds, this changes in the phase of constructing the phoneme system. That phase concentrates on selected sounds. This selection does not proceed randomly but according to regularity striving toward generality. On explanation for that regularity is the »principle of maximum contrast«:

In the first discrimination phase, if the tense bilabial plosive /p/ is set against the open velar long monophtong /a:/ (/pa:pa:/), these differentiations follow in the consonant system oral vs. nasal (/p/ vs. /m/) (/pa:pa:/ vs. /ma:ma:/), then labial vs. dental (/p/ vs. /t/ or /m/ vs. /n/) (/pa:pa:/ vs. /ta:ta:/, /ma:ma:/ vs. /na:na:/) etc. (also called »minimal consonantism«).

In the vowel system, the first distinction is between the open velar long vowel vs. the almost closed palatal monophtong (/a:/ vs. /I/) (/pa:pa:/ vs. /pIpI/), then open vs. half-open or half-open vs. closed (/a:/ vs. /e:/ or /e:/ vs. /i:/) (/papa/ vs. /pe:pe:/, vs. /pi:pi:/) etc. (also called »minimal vocalism«). The principles of minimal consonantism and minimal vocalism are illustrated in Fig. 11.1.

Although Jakobson's findings are still under discussion today this diachronic comparison hints at »structural principles« that guide the development of the phoneme system or speech system mainly in regard to speech production. It is these structural principles that are of main importance here. Of interest is Jakobson's realization that these structures are relatively stable. This is not an unproven assumption, but a conclusion derived from observations of gradual speech loss ("That which is last acquired is lost first of all.") or from known facts of phylogeny of language.

This short digression on the diachronic shows that speech acquisition processes do not occur freely, but that speech is learned systematically. Therefore, language is described as a system, even for rather varied aspects such as grammar or meaning. The concept »system« is understood to be a set of elements that show certain characteristics and that stand in relation to one another. Systems are characterized by a structural order (cf. [229], [254]). In this sense, Nöth describes language as a system:

> "Language is regarded as a system [...] of *elements* that stand in certain *relations* to one another. The elements and their relationships to one another form the *structure* of the system. Hierarchy, dependence, dominance, opposition, complementarity, class and distribution are among the essential concepts with which the structure of the speech sign is described in structural linguistics." [189]

In addition, Jakobson's diachronic examination of speech sound production suggests viewing the development of the perception of speech as a system as well. If speech perception has systemic character, it is thus presumably characterized by hierarchy, dependence, dominance, opposition, complementarity, class and distribution as well.

If this were proved, it would have consequences for the development of speech synthesis systems: Based on the system specifications, it would follow that it in no way depends solely on forming each individual element based on an isolated theoretical reference target value. Rather the goal would have to be to form each individual element in such a way that the relationship network of phonetical similarities between all elements is in accordance with that of natural speech. It is about the markedness and valency of each element that for its part is qualified by the structure of sound relationships.

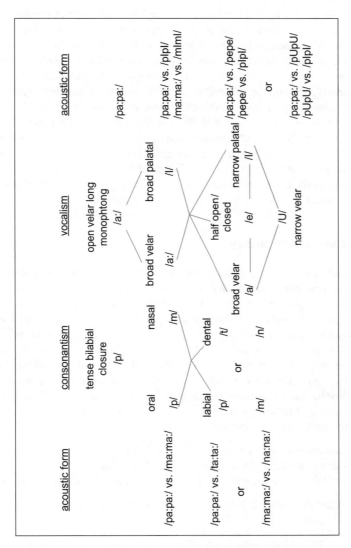

Fig. 11.1. Maximal contrasting as a structuring principle in language acquisition

The cluster similarity study attempts to find out more about conditions of elements, relations and structures of speech signal and auditory speech events. It is a contribution to pure research in the field of speech-related psycho-acoustics, initiated by fundamental questions and problems from the field of metrology and speech technology, in this case speech synthesis development in particular.

The study is based on the following model: Elements of auditory events (whether individual sounds, syllables, words, etc.) are understood as knots in a broad net. This net is imagined to be knotted of flexible threads. If the structure is changed at one point, it effects not only the area around it, but the whole net (this image is taken from [253]). Accordingly, to view speech as a system, not only isolated events, but also relationships between events must be understood. In the case of auditory perception, besides the elements themselves, the linkage plays an essential role. It is the goal of this study to examine this relation in greater detail.

11.2 Details of the study

It is presumed that the listener uses a network of entities which is characterized by hierarchy, dependence, dominance, opposition, complementarity, class and distribution. This is supported by the fact that there are speech entities (e.g. individual speech sounds) that sound similar (/b/ and /p/), and there are others that are easily distinguishable from one another (/b/ and /s/). The goal of this study is to portray, through listening experiments, areas of similarity of speech entities of natural and synthetic speech. For the preparation of the material to be investigated pre- and post-nuclear consonant clusters were selected as objects of investigation. All pre-nuclear consonant clusters were linked to the vowels |a|, |a:|, |aU|, |U| and |u:|, so that items such as the following were formed: |ta|, |pa|, |pfa|, |StRa|, ... (pre-nuclear list) and |at|, |ap|, |apf|, |ax|, ... (post-nuclear list).

- pre-nuclear consonant clusters: |k|, |pR|... (62 entries)
- post-nuclear consonant clusters: |k|, |x| ... (80 entries)
- vowel list: four monophtongs |a|, |a:|, |U|, and |u:| as well as the diphtong |aU|

For each of the three cluster lists and for each vowel, a separate inventory was produced. For example:

vowel |a|	pre-nuclear consonant cluster:	|ka|, |pRa|, ...
vowel |a:|	pre-nuclear consonant cluster:	|ka:|, |pRa:|, ...
vowel |U|	pre-nuclear consonant cluster:	|kU|, |pRU|, ...
diphtong |aU|	pre-nuclear consonant cluster:	|kaU|, |pRaU|, ...
vowel |u:|	pre-nuclear consonant cluster:	|ku:|, |pRu:|, ...
vowel |a|	post-nuclear consonant cluster:	|ak|, |ax|, ...
vowel ...		

Then each entry was embedded in a carrier sentence. There was one sentence per position:

pattern: »Das wäre   (pre-nuclear C-cluster+vowel)   telei gemacht.«
example: »Das wäre............................ **ta** .... telei gemacht.«

|das v'E:R@ .... **ta** .... t@'laI g@'maxt|

pattern: »Das Stoßgeb   (vowel+post-nuclear C-cluster)   ist ohne Sinn.«
example: »Das Stoßgeb .... **at** .... ist ohne Sinn.«

|das Sto:sg@b...................... **at** .... ?Ist ?o:n@ zIn|

Two studies were carried out: For study A the sentence material was read aloud by a professional speaker in an anechoic chamber, and for study B the same material was produced by synthesis A. The speech material was digitally recorded. Within the signal files the target stimuli were marked, cut out of the respective sentence with a signal editor and stored as individual stimuli.

For pre- and post-nuclear position as well as for vowels each, lists of paired syllables were produced in which each entry was permuted with each other entry:

|ta| vs. |pa|, |ta| vs. |StRa|, |ta| vs. |kva|, |ta| vs. |pRa|, |ta| vs. |pla|, ...

The perceived similarities of each pair should be scaled. For this purpose, however, the stimuli are paired in one-sided permutations only. In this way, /ta/ vs. /pa/ is tested, but not /pa/ vs. /ta/. Each pair is transmitted one time only. This is due to the vast amount of stimuli. Per vowel, there are 1953 stimulus pairs for pre-nuclear consonant clusters and 3240 stimulus pairs for post-nuclear consonant clusters. If the number of pairs is then multiplied with the 5 selected vowels, the total number of stimuli per test person is 25.965. Due to the extremely high number of stimuli (they are still to be multiplied by 2, since synthetic and natural speech components are to be compared), only 2 TPs ran the entire investigation. In order to check the reliability of results, a subset of stimuli was tested with 15 TPs.

The test person is presented the stimulus pair acoustically (e.g. /ta/ vs. /va/). The task is to mark the perceived similarity on a 5-point scale (1=extremely similar, 5=extremely dissimilar). In the introductory phase the TPs were instructed to use the scale in intervals, i.e. to assign the numbers assuming equidistant intervals between them.

11.3 On rating and scaling

The main part of the study was carried out with 2 test persons. TP1 is a female student, 24 years old, with normal hearing capacity, TP2 is a male student, 27 years old, normal hearing capacity. It shows that both TPs have an extremely similar manner of rating (cf. Table 11.1 and Table 11.2). The frequency of the scale points indicated for the individual stimuli is represented. The distribution makes a succinct pattern.

If the individual ratings are viewed, it is presumed that, with relatively high probability, they are in fact assigned the perceived similarity characteristics of the voice (the measuring object). There is no indication at this point that these two TPs have not made use of the scale uniformly. Both distributions are sloping on the left and steep on the right. Of note is that the increase from point 4 to point 5 is steep in all cases. Both TPs used all available points of the scale. Based on these data, it can be assumed first of all that the scale offers a suitable form to take down the feature 'phonetic similarity'.

If a pilot study is carried out with only two test persons, assumptions can only be made with restraint based on the results. However, in view of the immense number of stimuli to be assessed and the limited resources available, it was not possible to involve a greater number of test persons in the full pilot study. In spite of that, in order to at least be able to estimate roughly the performance of these two test persons, a control group of 13 test persons was formed (all with normal hearing, between 21 and 33 years old, students). This control group judged a subset of all stimuli. It is the set of pre-nuclear consonant clusters, linked with the vowel |a| for natural and synthetic voices.

Table 11.1 and Table 11.2 graphically represent the discerning performance of the entire TP group for the same set of stimuli. These data are clearly more spread out, as expected, as those from TP1 and 2. Of note are TPs 5, 7, 10, 11 and 14: TP 5 apparently evaluates the similarity in a much more subtly differentiated manner than all other TPs and as a result uses the entire spectrum of the scale extensively, while TPs 7, 10, 11 and 14 apparently hesitate more frequently in using the extreme values 1 and 5. This is a well-known problem in psychometry. It can be traced back to the closed nature of scales. TPs are shy of employing extreme values because they do not know what quantity and quality the subsequent stimuli will have. To hold open the possibility of marking on the scale even more extreme characteristics than those perceived up to that point, the neighboring values (2 and 4) are preferred in such cases. The consequence is that a Gaussian distribution can no longer be assumed.

The normal distribution is a standardized Gaussian distribution, in which the most values are grouped around the middle point and the frequency density falls comparably to both sides. Depending on whether or not a normal distribution is given, so-called parametric and non-parametric tests are used. Among the parametric tests are e.g. the t-Test by Student, and among the non-parametric are the U-Test by Mann and Whitney as well as the Kolmogorov-Smirnov Test (cf. [101]).

If no Gaussian distribution results, it can be useful to transform the scale. In this way the data distribution is evened out. However, the precondition is that distribution prominence can be clearly traced back to a scaling of intervals possible in principle, but in fact missing. This, however, is not proved here for the cluster similarity study. It should be noted that the outliers described are regarding 5 of 15 TPs. For all other, especially TP 1 and 2, a scaling of intervals possible in principle can clearly be assumed.

This discussion makes clear that the results are to be analyzed and interpreted with care. However, that by no means indicates that the data are not useful. On the contrary, a comparison of results of both TPs who took part in the full study with those who processed a subset of these stimuli shows that the subset is rated similarly to the average rating. This increases the assumed reliability of the results.

As already said, it could not be the goal from the beginning to collect representative results for a greater group of listeners. For that, the stimulus material is too extensive, the time needed too great and the costs too high. The goal was to put together a stimulus set that is representative with respect to defined features, despite the variability of the phonotactics in general. In our case here defined features are

Table 11.1 Natural voice: use of scale per TP

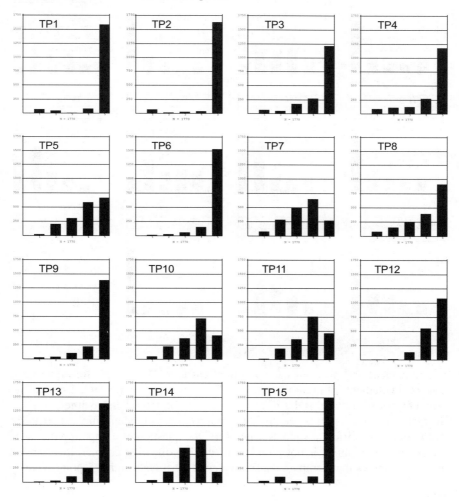

Table 11.2 Synthetic voice: use of the scale per TP

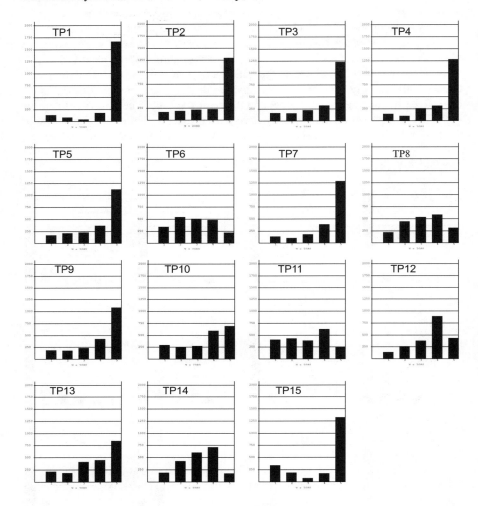

the pre-nuclear and post-nuclear consonant clusters. The vowels linked to these clusters are elements of minimal vocalism. Then individually perceived similarities should be portrayed for this defined set of stimuli, for natural and synthetic speech. The measuring goal is consequently the comparison of individual similarity patterns of stimuli of two different sources. Therefore, the question of computing the mean or median, for example, is not posed at all. In the following chapter, there are, in fact, two graphics that compare the frequency of portrayals on the 5-point scale by the 15 TPs with respect to natural and synthetic voices. They are, however, solely meant to depict general tendencies. It would be inappropriate to interpret more than that into the data.

11.4 Natural voice: Similarity profile of pre-nuclear consonant clusters

In Fig. 11.2., the individual results from TP 1 and TP 2 for the natural voice (consonant cluster + /a/) are shown in a matrix – the results for TP 1 in the left, upper half, and from TP 2 in the right, lower half. The graphic can be read as follows:

Both coordinates list the entries of the stimulus inventory. The grey scaling of the intersection shows the degree of perceived similarity of a stimulus pair. White means »dissimilar«, black »similar« and the three variations of gray the corresponding values in between. Uniformity (e.g. /ta/ vs. /ta/) was not questioned. Therefore, the diagonal remains open (marked in white).

For TP 1, the sphere of similarity for natural speech is occupied only at a few points by stimuli that are judged as phonetically very similar. A closer analysis of this point indicates three dominant phones:

- the /s/ in the contexts /skl/, /sl/, /sR/, /ps/, /sts/, /ks/ and /sk/
- the /S/ in the context of /C/, /SR/, /St/, /Sl/, /Sv/, /Sp/, /SpR/, /StR/ and /Spl/
- the /f/ in the context of /pf/, /fR/, /pfR/ and /pfl/

In these clusters, the /s/, /S/ and /f/ appear to perceptually dominate all further consonants that make up the respective cluster. The same tendency is indicated by the hierarchical cluster analysis. The hierarchical cluster analysis is a statistical procedure that groups cases generally based on predefined variables. The concept »cluster« as used here has, of course, another meaning than the »cluster« in the context of cluster similarity study: In the field of statistics the concept »cluster« describes such groups that are formed in the process of classifying results.

Hierarchical cluster analysis takes place in two steps: with the selection of the proximity measurement and with the selection of the fusion algorithm. As explained on page 150, it can be assumed that the test results of both TPs are scaled by intervals. For this reason, the quadratic Euclidean distance can be used as proximity measurement. The method used is the Ward Method. It is a hierarchical agglomerative procedure. The steps in procedure are briefly described as follows:

- First of all, each object (in the case of the similarity study each consonant cluster) represents a separate cluster (fine partitioning).
- A distance measurement is calculated for all objects (here the Euclidean distance).
- The clusters that have the slightest distance from one another are marked and combined in a group.
- The distance between the groups (clusters) are calculated again (reduced distance matrix).

These steps are repeated until all clusters are combined in a group. The goal of the Ward Method is to unite those objects that raise the variance in a group as little as possible. As homogenous a group as possible is formed. Objects that increase a pre-defined heterogeneity measurement the least are combined. The variation criteria (error square sum) is used as the heterogeneity measurement. (More details on the hierarchical cluster analysis and the Ward Method cf. [7]).

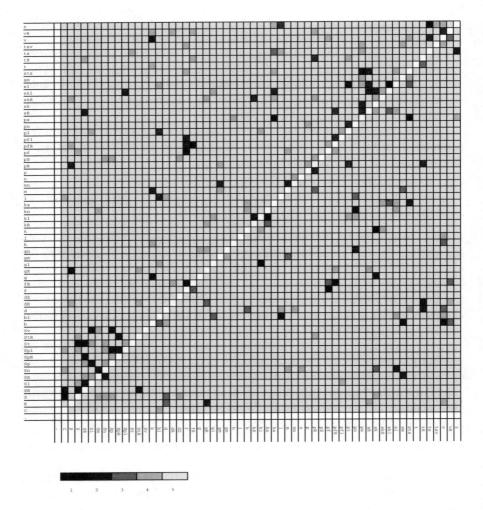

Fig. 11.2. TP1 (left upper half) and TP2 (right lower half), natural voice, pre-nuclear consonant clusters, vowel /a/

The so-called »dendrogram« is selected here to present the results in a representative way. The dendrogram gives a quick overview of the groups formed. In the selected Ward Method, the highest value corresponds to 25 on the scale of error square sums of the last fusion step.

The dendrogram in Fig. 11.3. represents the perceived similarity of consonant clusters by TP1, the one in Fig. 11.4. is related to TP2. It clarifies quite well that the similarity space is in fact characterized by opposition, complementarity and hierarchy. As expected, not all consonant clusters are classified as having the same degree of similarity or dissimilarity. There are in fact groupings that result from the individual fusion steps and that extend relatively evenly over the entire similarity space.

The dendrograms represent the perceived similarity of consonant clusters. The data material can then be used as a model to explain the process of identification of speech sound sequences. In the model, the direction of fusion is not examined (the formation of classes), but the opposed direction of differentiation or classification in the sense of Jakobson:

When a listener perceives the speech sound items studied here, the sounds are not classified by the entire feature substance, i.e. by all perceivable components, but by contrasting prominent individual features. A mostly binary decision tree is run for classification (with respect to the dendrogram that is from the highest point of 25 on the scale in the direction of 0, i.e. from the last fusion step to the first – which means from the first classification step to the last). The trend of the classification indicates the role of perceivable speech sound features, i.e. the morphological and syntactic-semantic comparison is faded out. These study conditions assure that the perception of speech sound segments is based not on individual sounds, but on auditorily prominent characteristics of items and their relations.

Comparing the results from TP1 and 2, it stands out that TP2 extremely infrequently evaluates very fine similarity features on the lowest level of identification (which is the first fusion step). The only very similar pairs are /ts/ and /z/, /SR/ and /StR/, /b/ and /v/, /f/ and /pf/ as well as /fR/ and /pfR/. All other clusters are already grouped on higher levels (not on the level of the finest sound contrasting). A prominent group is the clusters /j/, /tsv/, /h/, /Sv/, /pS/, /SpR/, /Spl/, /St/, /n/, /mn/, /C/, and /kR/.

In comparison to TP1, for TP2 other features apparently dominate. In fact, there is also the /f/– group with the consonant clusters /f/, /pf/, /fR/ and /pfR/, but there is just as little evidence of a distinct /S/–group as the noticeable /s/–group for TP1. Instead the phones /l/, /m/ and /n/ perform a superior function in the classification process.

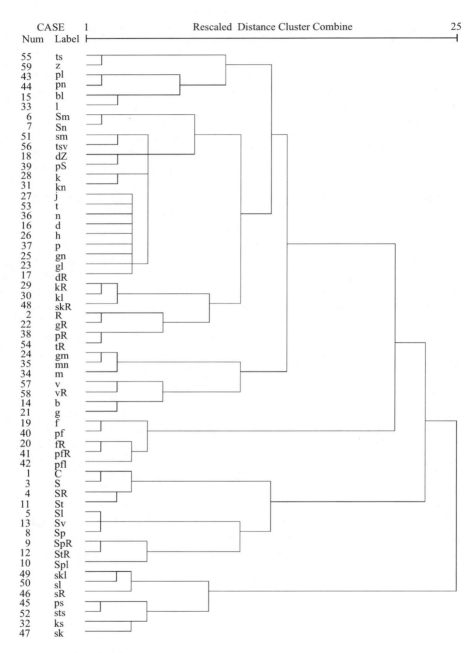

Fig. 11.3. TP1. Natural voice, pre-nuclear consonant clusters, vowel /a/

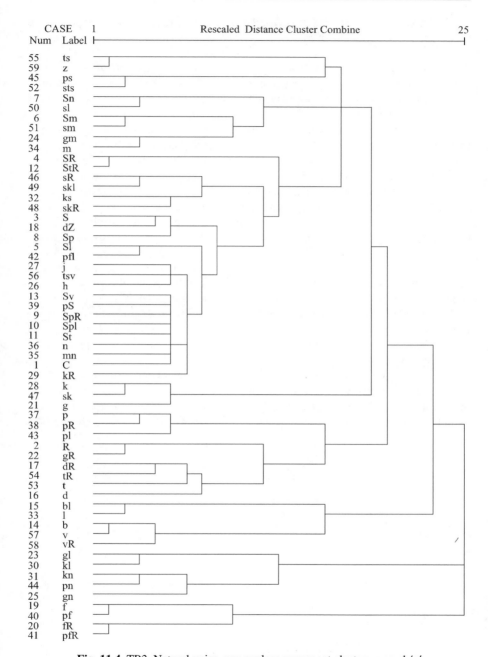

Fig. 11.4. TP2. Natural voice, pre-nuclear consonant clusters, vowel /a/

11.5 Comparison of the similarity profile of natural and synthetic speech

As mentioned previously, the cluster similarity study was carried out under the same conditions for a natural and for a synthetic voice. The synthesis chosen is system A, used in the previous chapters already. The results from TP 1 and 2, for pre-nuclear consonant clusters, synthetic voice, vowel /a/, are shown in Fig. 11.5. In contrast to natural voice the similarity profile of both TPs is restless and unstructured for synthetic voice. That is also verified by the comparison of corresponding dendrograms from TP 1 (cf. Fig. 11.3. on page 156 and Fig. 11.6. on page 160) and TP2 (cf. Fig. 11.4. on page 157 and Fig. 11.7. on page 161):

In contrast to the natural voice, many objects in the synthetic voice are classified first in the lowest level, whereby the corresponding groups contain up to seven objects. This trend in the assessment of the synthetic voice is not as prominent for TP 1 as for TP2: After the first fusion step of the hierarchical cluster analysis, two groups of seven objects, two groups of five, one of four, eight of three and only two groups of two objects were formed from the similarity portray given by TP2. In order to clarify in detail the difference, the data for the natural and synthetic voices are presented together in Table 11.3 The table shows that, for the natural voice, distinctive features that aid in contrasting are used for classification on a high level. The respective groups after the first fusion are very small for the measuring object »natural voice«. This is a further indication of the effectiveness of Jakobson's principle of maximal contrasting.

The similarity assessments for the synthetic voice stand out very markedly from this picture. Here there are fewer classification levels and the number of objects in the groups is significantly greater. This may be used as an explanation, supported by data, why TPs do not understand synthetic speech as well as natural speech. The

Table 11.3 Number of object groups for natural and synthetic voice for TP 1 and TP2 (after the first fusion step)

		groups à x objects on the last classification level					
		7	6	5	4	3	2
TP1	natural voice					1	12
	synthetic voice	1			2	5	12
TP2	natural voice						5
	synthetic voice	2	1	1	1	8	2

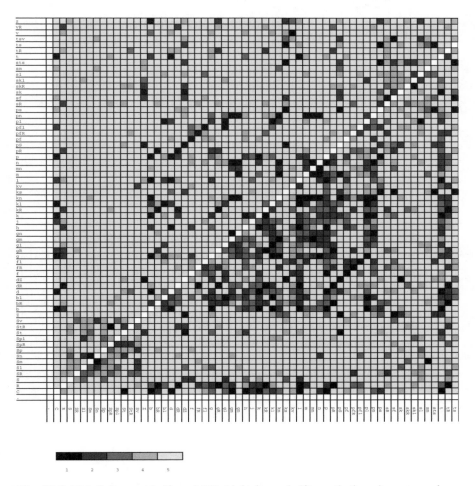

Fig. 11.5. TP1 (left upper half) and TP2 (right lower half), synthetic voice, pre-nuclear consonant cluster, vowel /a/

consequence is that the so-called »ease of communication« is impaired. The cause of this could be that the individual objects (the consonant clusters) do not show *the* distinctive features on which the reception apparatus is optimized.

Analyzing then the results for the natural and the synthetic voice for all 15 TPs as shown in Fig. 11.8. (as mentioned with reservations) the following is recognized despite all diversity and individuality:

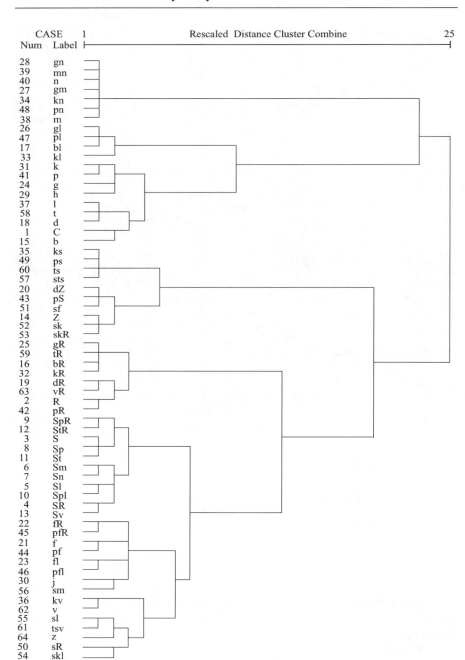

Fig. 11.6. TP1. Synthetic voice, pre-nuclear consonant clusters, vowel /a/

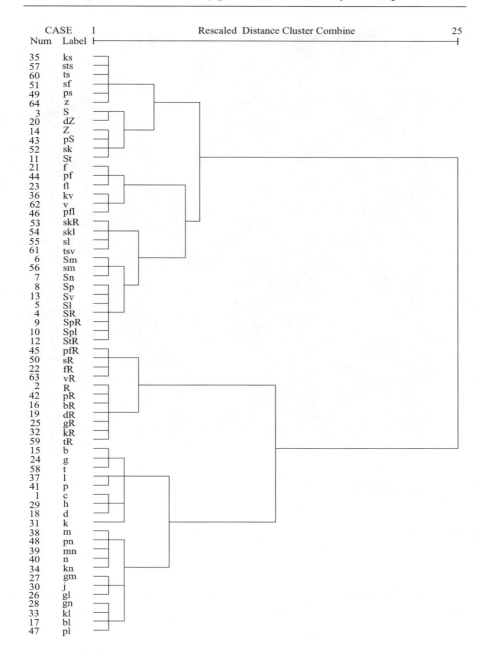

Fig. 11.7. TP2. Synthetic voice, pre-nuclear consonant clusters, vowel /a/

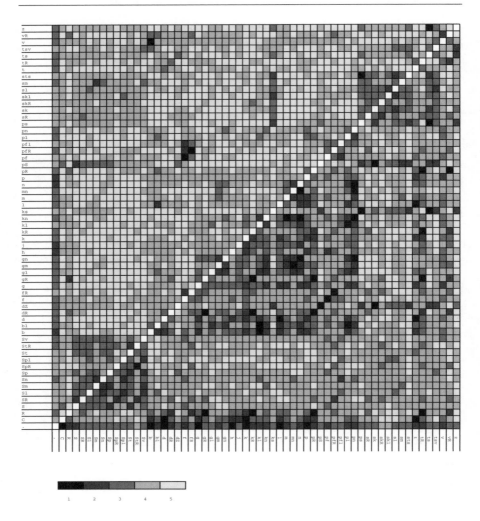

Fig. 11.8. 15 TPs. Natural voice (left upper half) and synthetic voice (right lower half), pre-nuclear consonant clusters, vowel /a/

For the natural voice, exceptionally few areas with a high degree of similarity in sound are formed. The areas that form, however, are found again in the result material of the synthetic voice. In addition, the similarity profile of the synthetic voice stands out in regard to a distinct cloud of perceived similarity in the middle of the matrix. An outline of this cloud is indicated in the profile of the natural voice as well.

It follows that these data also support the assumption formulated above that the synthetic voice does not carry the signal characteristics that listeners of speech normally use to contrast, distinguish and identify. This is confirmed as well by the comparative analysis of the dendrograms and further analytic data (cf. Fig. 11.9., Fig. 11.10., and Table 11.4).

The perceived form of natural voice is characterized by a hierarchical structure that, on the basis of perceived, complementary features, leads to a systematic classification of auditory objects. The individual objects of natural speech show noticeable similarity features that concern not only to the fine structure of the form. The principle of maximal contrasting described by Jakobson can be very clearly confirmed by the result material of the cluster similarity study. That means that the contrasting mechanism in speech production is indeed found here on the receptive side.

The comparison with the results for synthetic voice shows that the learned principles for maximal contrasting are not automatically applicable to artificial voices. Artificial voices show features that natural speech apparently does not possess. The study has shown that clearly. However, that alone has been proved. The question remains open whether synthetic speech signals possess no distinguishable characteristics at all or whether they are, in fact, characterized by features which do not reach the attention of the listener. In the process of language development the listener has learned to concentrate on those features that are communicatively informative. The reference object in the learning phase is natural speech.

In the case of synthetic speech, this is suddenly different: Artificial acoustic signals *should* carry information, and the form of the information carrier attaches to the known form of natural speech. That means that the receptive manner of the listener is directed, first of all, not at a new form, but at the processing of an acoustic information carrier, therefore of natural speech. The consequence is that synthetic signals are assimilated on the ground of the fully differentiated and set schematas of natural speech. The similarity profiles verify this at a further point:

Table 11.4 Number of object groups for natural and synthetic voice for 15 TPs (after the first fusion step)

	groups à x objects on the last classification level					
	7	6	5	4	3	2
natural voice, 15 TPs						8
synthetic voice, 15 TPs		1	1	2	7	7

As mentioned previously, the study was carried out not only for the stimulus material described up to this point, but also for additional consonant–vowel combinations. Analysis of these data shows that synthetic speech signals indeed possess perceivable features for the purpose of contrasting sounds. In the phase in which the listener is confronted with synthetic speech the first time, these features are too foreign and are, therefore, not used for an easy identification. In time, the listeners in fact accommodate to the new auditory object. This learning effect is, with time, also present in the stimulus material studied.

11.6 Similarity profile for various vowel contexts

Considering the similarity profile for the natural voice per TP for various vowel contexts, it stands out that TP 1 always makes use of quite similar structuring. The tendency is confirmed that oppositions and classifications are formed on a high level of contrasting (cf. Fig. 11.9. on page 167 and Fig. 11.10. on page 168).

TP 2 shows a quite noticeable manner of classification for the stimuli in the vowel context /a:/ and /aU/ (cf. Fig. 11.11. on page 169 and Fig. 11.12. on page 170). There is confirmed the trend that consonant clusters are organized with respect to their perceived similarity and that many items are already classified as distinct on higher classification levels. Contrary to previous observations and assumptions, the number of elements in the clusters is conspicuously large. Thus one cluster contains 37 elements (natural voice, pre-nuclear consonant cluster, vowel /a:/, TP 2); these are, however, all endpoints of a distinguishing level. That means that 37 different features are necessary for classification at this intersection. This trend is also found in a somewhat weaker form in the assessments of TP 1.

What is of interest to analyze is that this rating almost exclusively appears in vowel context /a:/ and /aU/. In the vowel context /U/ and /u:/ (cf. Fig. 11.13. on page 171 and Fig. 11.14. on page 172) a grouping pattern as in the vowel context /a/ dominates once more. Here the number of objects grouped together is again very small (mostly 2), and differentiation takes place in practically even steps over the entire similarity space. This is equally true for TP 1 and 2.

The question has to be answered whether these findings are really in the material or whether they are due to measuring errors. One cause of a measuring error could be due to the manner in which the material is spoken, another to the speech material (e.g. coarticulation effects) and, naturally, also to the rating and scaling of the auditory events. The last explanation, however, seems rather improbable because the rating of both TPs is quite similar. Furthermore, the sequence the studies were carried out were first /a/, then /a:/, /aU/, /U/ and /u:/ context – concentration difficulties or training effects can be ruled out in all probability since the noticeability would be expected toward the end of the questioning and not in the 2nd and 3rd series of experiments. The study conditions remain the same during the entire study.

As previously described, the similarity profiles for synthetic speech conform increasingly to those of the natural voice in the course of the study. That shows that both TPs open themselves to demands of the new perceptive object and make use of features unconsidered or differently weighted up to that point. That leads in the end to a remarkable alignment of the structure being achieved, with respect to groupings and hierarchizing of items (cf. Fig. 11.14. on page 172 and Fig. 11.15. on page 173). This can also be observed in the post-nuclear clusters.

11.7 Summary

• The cluster similarity study is targeted at form characteristics of voice. Aspects of content are not considered. The paradigm chosen is monosyllables.

• Carrier sentences of typical syllable-structures were used as syntagma, in which each time meaningless monosyllables were substituted paradigmatically. The monosyllables were scaled with respect to the perceived similarity.

• The goal was to gather information regarding the number of weighted similarity each consonant pair is characterized by and how the similarity space can be established. It was shown that similarity spaces of different listeners indicated an analog, categorical structure – which does not mean that it is also isomorphically structured.

• A further goal was to use the procedure to calculate the degree of difficulty of speech comprehensibility in connection to further information-based theoretical approaches, for instance, from the statistical similarity value of speech sounds in correlation with their respective textual occurrence values. This goal lies far in the future because, among other things, this procedure requires an immense compilation of data, and the means to achieve this have not been available up to the present. The data compiled in this study alone are extremely extensive. For this reason concentrating on the selected aspects was necessary.

• The study should, therefore, be classified as a feasibility study. It was carried out to find out whether the measurement procedure is a suitable means to extend the general knowledge with respect to perceived form characteristics of speech sounds. In this sense, it achieved its goal.

• The results show that there are clear similarity spaces of perceived speech sounds (here only consonant clusters). As mentioned repeatedly, the quality of speech results from the expectation of the listener, among other things. The results of the cluster similarity study quantify for the first time listener-specific expectations for voices. This is certainly only for an exactly defined section in which speech and speech perception can take place; it is, however, informative in many regards. Furthermore, natural and synthetic speech are compared to one another for the first time. In addition, the procedure demonstrates listener-specific assimilation and accommodation behaviors and verifies – again, quantitatively – both that, with time, listeners adapt to synthetic speech sounds and it records the manner in which they adapt.

• Consequently, the results of the cluster similarity study are in accordance with the hypothesis that the valency of individual speech components is formed through contrasting. That explains why humans are able to understand quite different ways of speaking after a relatively short period of exposure. It explains as well why humans have difficulties perceiving the form of the synthetic voice as pleasant. Based on the data analyzed, the following concept is supported: Speech components are knots in a network and the value of the knots is determined by the relation to other knots. A feature modification in one knot leads to changes in relation to and thus the value of all other knots, especially the ones in the immediate surroundings.

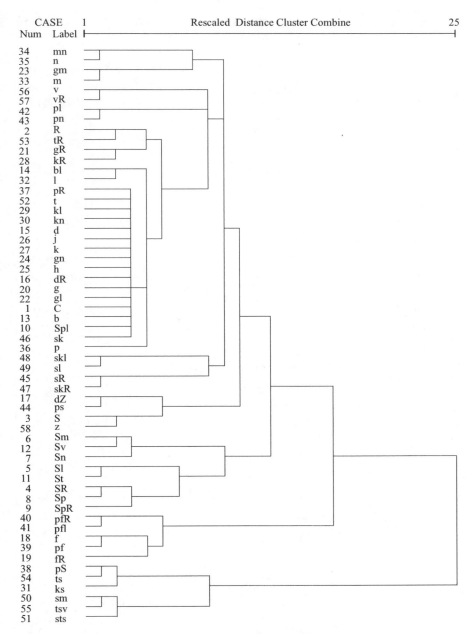

Fig. 11.9. TP1. Natural voice, pre-nuclear consonant clusters, vowel /aU/

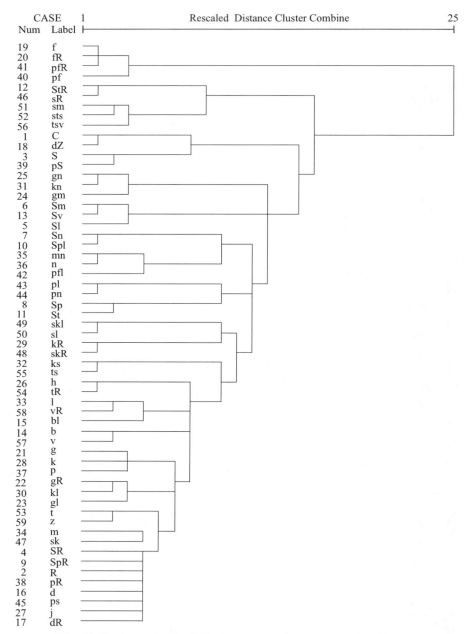

Fig. 11.10. TP1. Natural voice, pre-nuclear consonant clusters, vowel /U/

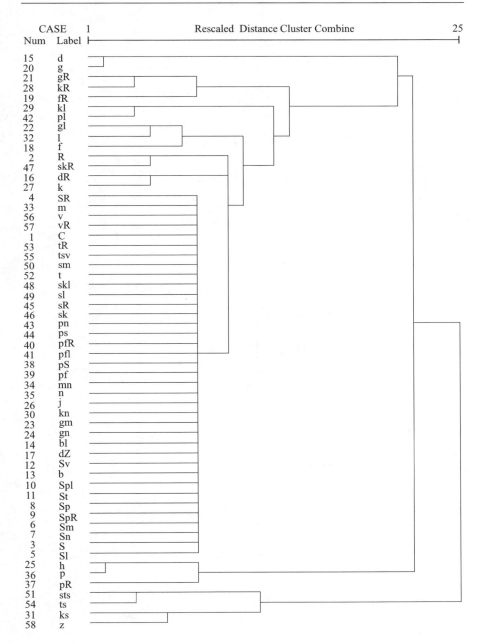

Fig. 11.11. TP2. Natural voice, pre-nuclear consonant clusters, vowel /a:/

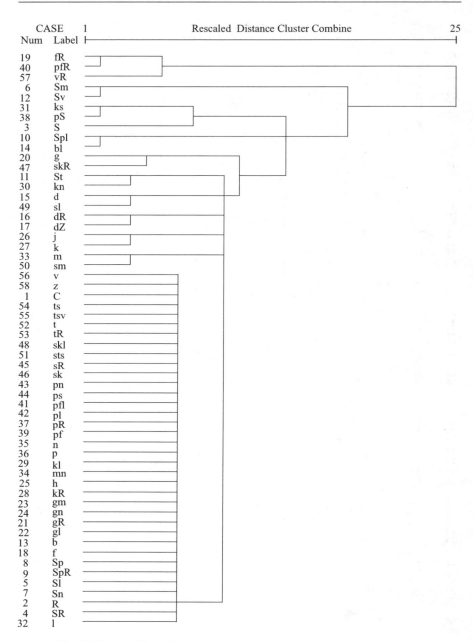

Fig. 11.12. TP2. Natural voice, pre-nuclear consonant clusters, vowel /aU/

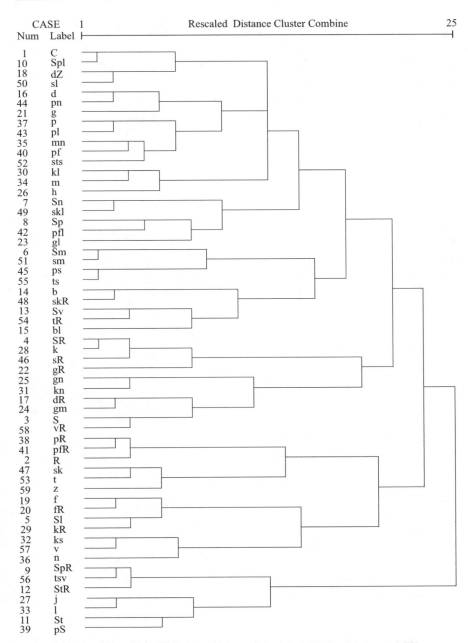

Fig. 11.13. TP2. Natural voice, pre-nuclear consonant clusters, vowel /U/

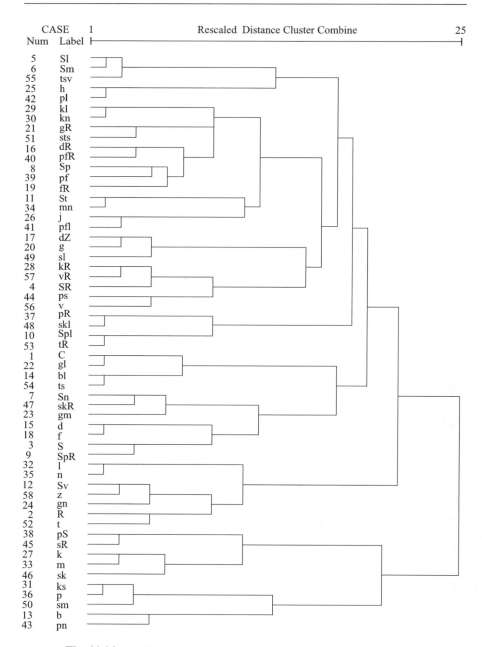

Fig. 11.14. TP2. Natural voice, pre-nuclear consonant clusters, vowel /u:/

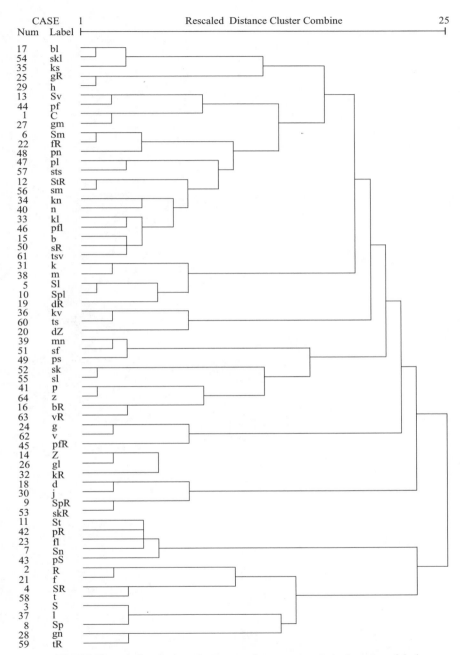

Fig. 11.15. TP2. Synthetic voice, pre-nuclear consonant clusters, vowel /u:/

12 Conclusion

In this book we have approached the issue of voice and speech quality perception as a research topic, starting out with the fundamental question of:

> *How do listeners perceive voice and speech quality and how can these processes be modeled?*

Any quantitative answers to this question require measurements. The science that deals with measurements and scaling is metrology, whereby, from a classical point of view, metrology is understood as being directed toward physical measurands exclusively. With regard to the topic of this book, such a restriction would, however, imply that acoustic speech events would be measurable, but not their perceptual counterparts, namely, the perceived speech sounds. Implicitly, this would also mean that any assessment of perceptual objects by means of human observers (subjects) would fail as a measurement, although quantitative data may result from it.

One of the reasons for the reluctance to accept dimensions of perceptive quality events as measurands is the fact that there is no basis of reference in the following sense: Measuring is a planned activity in which a quantitative comparison of a measurand and a reference quantity of the same dimension – a unit – is carried out. This procedure, which is obvious for physical quantities, is harder to imagine for perceptual measurands, as the relevant dimensions, and even more so, appropriate units have often not yet been identified and/or defined. Consequently, as long as we do not know the relevant dimensions of speech quality and as long as we do not have defined units respectively, we cannot measure. The only possible way out of this dilemma is to actually identify major perceptual dimensions of voice and speech quality perception, define units wherever possible and to offer paradigms to position these dimensions into a structural skeleton of perceptual speech and voice quality.

This, indeed, is the approach as taken in this book. Starting off from a theoretical perspective, the field of voice and speech quality perception is first examined as an integrative structural concept. These conceptual considerations result in what we call a »projection model of voice and speech quality«. Elements of measurements are understood as knots in a broad net. This net is imagined to be knotted of flexible threads. If the structure is changed at one point, it effects not only the area around it, but the whole net. Accordingly, to view speech quality perception as a system, not only isolated events, but also relationships between events have to be understood. In other words, besides the elements themselves, the linkage plays an essential role. Hierarchy, dependence, dominance, opposition, complementarity, class and distribution are most important cues. This perspective motivates the basic concept of a projection model as a reference. All the specifications which go into the projection model are exclusively motivated by auditory perception. Particularly, they are neither based on *aspects of the language system*, nor on *acoustic classification schemes* or on *theoretic approaches*. What is in focus is perceived speech quality, and from this point we go back and identify the relevant quality dimensions which are required by metrology. Using a term of phonetics, we determine »distinctive features« of voice and speech quality perception in a »minimal pair analysis«. One might also call this an approach toward the identification of »testemes« of voice and speech quality measurements.

The theoretical considerations plus a number verifying experiments result in the following statements:

• There is no justification to principally reject auditory speech quality tests as being »subjective« and to refuse to classify them as measurements. The listeners' behavior cannot be reduced to arbitrariness, rather it is strategic and subject to a certain systematology. It is our task to understand this systematology. Based on a systemic ground, auditory assessment tests can be constructed and performed in such a way that they can be classified as measurements. The same holds for observation as well as for calculating and statistical estimating processes.

• In order to classify events of voice and speech quality perception as measuring processes, a number of requirements must be fulfilled. Among other things, the measuring principle and the measuring method must be clearly described, and ongoing processes must be controllable. However, understanding the assessment processes alone is not enough to provide a sufficient basis for claiming to measure, one has also to be aware of what one is leaving out. For this purpose the projection model may serve as a protocoling editor.

• Further, we talk about measurement only when our approach is scientifically substantiated, i.e. when it can be carried out routinely under standard conditions, and when it allows a relative definition of position of the object to be measured with respect to other objects or other measurands on a scale. If one of these requirements is not fulfilled, we do not speak about a measurement any more but rather about an »investigative process« or an »experiment«.

• One aim of current activities in the field of speech quality assessment and evaluation is to develop instrumental procedures to either rate, monitor or even predict speech quality. Independent of what the ultimate objective is, it will rely on reports on auditory voice and speech quality perceptions. These reports, in turn, are dependent upon whether data on perceived speech quality have quality themselves, i.e. whether they are measured values in the sense described above – or simply estimations or guesses. In our context this means, among other issues, that measurements of perceived voice and speech quality are used as background of ongoing research and development.

In this book, voice and speech quality have been approached in the isolated context of speech stimulus and quality perception, mainly focussing on form characteristics alone. This was necessary because the objective was to introduce and explain the fundamental approach by referring to some experimental data. The scope taken taken into account is certainly only a small subset compared to how speech quality is perceived when it is processed as a sign carrier, i.e. a carrier of meaning. And it is an enormous simplification of how speech is actually perceived in natural communication environments. Further aspects that we have not dealt with but which are major objects of research today are, for example, the relationships between perceived speech quality and behavior in multimodal perceptual contexts. However, as already said a certain amount of simplification has been unavoidable here in order to be able to detect and define some major characteristics of voice and speech quality. We can, of course, not exclude that the ideas we have developed from this simplified approach may have to be modified in the course of further differentiation of the research area that we have dealt with in this volume.

13 References

[1] Adams E, Messik S (1963) An axiomatic formulation and generalization of successive intervals scaling. In: Luce R D, Bush R R, Galanter E (eds) Readings in mathematical psychology. Wiley & Sons, New York, pp 3–16

[2] Allen J, Hunnicutt M S, Klatt D (1987) From text to speech: The MITalk system. Cambridge UP, Cambridge

[3] American National Standards Institute S 3.5 (1997) Methods for the Calculation of the Speech Intelligibility Index (SII). ANSI, New York

[4] Argente J A (1992) From speech to speaking styles. In: Speech Com 11, pp 325–335, 328

[5] Aschoff V (1984) Geschichte der Nachrichtentechnik. (engl.: History of information technology) Springer, Berlin Heidelberg New York

[6] Ausubel D P, Sullivan E V (1978) Historischer Überblick über die theoretischen Ansätze. (engl.: Historic overview of theoretical approaches) In: Steiner G (ed) Die Psychologie des 20. Jahrhunderts. Bd. VII: Piaget und die Folgen. (engl.: The psychology of the 20th century. Vol. VII: Piaget and the consequences) Kindler, Zürich, pp 547–67, 553

[7] Backhaus K (1990) Multivariate Analysemethoden. (engl.: Multivariate analysis methods) Priner, Berlin

[8] Bappert V, Blauert J (1994) Auditory quality evaluation of speech-coding systems. In: acta acustica, 2, pp 49–58

[9] Beerends J G, Stemerdink J A (1994) A perceptual speech quality measure based on a psychoacoustic sound representation. In: J Audio Eng Soc 42(3), pp 115–123

[10] Belhoula A (1990) Algorithmische Beschreibung der Phonotaktik des Deutschen und Entwicklung eines Programmsystems zur Generierung von phonotaktisch korrekten sinnlosen Wörtern. (engl.: Algorithmic description of the phonotactics of German and development of a program system for the generation of phonotactically correct non-semantic bearing words) Dipl. thesis, Ruhr-University Bochum

[11] Bench J, Kowal A, Bamford J (1979) The BKB (Bamford-Kowal-Bench) sentence lists for partially-hearing children. In: Br J Audiol, 13, pp 108–112

[12] Bennet R W (1988) Assessing speech quality. Context effects in listening experiments. In: The Hague Human Factors in Telecommunications, Session D, Paper 2

[13] Benoit C, Van Erp A, Grice M, Hazan V, Jekosch U (1989) Multilingual synthesizer assessment using semantically unpredictable sentences. In: Proc Eurospeech, Paris, France, pp 633–636

[14] Berger J (1998) Instrumentelle Verfahren zur Sprachqualitätsschätzung – Modelle auditiver Tests. (engl.: Instrumental methods for speech quality estimation – models of auditory tests) Shaker, Aachen

[15] Blauert J (1967) Bemerkungen zur Theorie bewusst wahrnehmender Systeme. (engl.: Some comments on the theory of consciously perceiving systems) In: Bense M, et al (eds) Grundlagenstudien aus Kybernetik und Geisteswissenschaft. (engl.: Basic studies from cybernetics and the arts) vol 8/2, Schnelle, Quickborn, pp 45–56

[16] Blauert J (1969) Die Beschreibung von Hörversuchen anhand eines einfachen, systemtheoretischen Modells. (engl.: The description of auditory experiments by means of a simple system-theoretic model) In: Kybernetik 5, pp 45–49, 48

[17] Blauert J (1974) Akustik II: Kommunikationsakustik. Skript zur Vorlesung Akustik im Bereich allgemeine Elektrotechnik und Akustik an der Ruhr-Universität Bochum (engl.: Acoustics II: Communication acoustics. Accompanying script to the lecture series on acoustics in the field of electrical engineering and acoustics at Ruhr-University Bochum), Ruhr-University Bochum

[18] Blauert J (1996) Spatial Hearing. The Psychophysics of Human Sound Localization. rev edn MIT P, Cambridge, p 5, 8

[19] Blauert J, Bappert V (1990) Auditive Güteanalyse von Systemen zur kodierten Sprachsignalübertragung. (engl.: Auditory quality analysis of systems for coded speech signal transmission) (Final report by the Ruhr-University Bochum, Telecom Project)

[20] Blauert J, Jekosch U (1996) Sound-quality evaluation – a multi-layered problem. In: acta acustica, 83, pp 747–753

[21] Blauert J, Jekosch U (2003) Concepts of sound quality: some basic considerations. In: Proc Internoise, Jeju, Korea, pp 72–79

[22] Blauert J, Schaffert E (1985) Automatische Sprachein- u. -ausgabe. Verfahren, gebräuchliche Systeme, menschengerechte Gestaltung. (engl.: Automatic speech input and output. Methods, usable systems, human adequate design) Wirtschaftsverlag NW, Bremerhaven, pp 66–68

[23] Block von S (1998) Von der Systemlinguistik zur Patholinguistik. Interdisziplinäre Verflechtung anwendungsbezogener Forschung. (engl.: From system linguistics to patholinguistics. Interdisciplinary combination of applied research) Lang, Frankfurt aM, p 108

[24] Bodden M (1992) Binaurale Signalverarbeitung: Modellierung der Richtungserkennung und des Cocktail-Party-Effektes. (engl.: Binaural signal processing: modeling of directional hearing and the cocktail-party-effect) VDI-Verlag, Düsseldorf

[25] Bodden M, Jekosch U (1996) Entwicklung und Durchführung von Tests mit Versuchspersonen zur Verifizierung von Modellen zur Berechnung der Sprachübertragungsqualität. (engl.: Developing and performing auditory experiments for verifying speech transmission quality prediction models) (Final report by the Ruhr-University Bochum, Telecom Project)

[26] Boogaart T, Silverman K (1992) Evaluating the overall comprehensibility of speech synthesizers. In: Proc Int Conf on Spoken Lang Proc, ICSLP '92, Banff, Canada, pp 1207–1210

[27] Borg G (1982) A category scale with ratio properties for intermodal and inter-individual comaprisons. In: Geissler H G, Petzold P (eds) Psychophysical judgment and the process of perception. Deutscher Verlag der Wissenschaften, Berlin, pp 25–34

[28] Borg G (1994) Psychophysical scaling: an overview. In: Boivie J, Hansson P, Lindblom U (eds) Touch, temperature, and pain in health and disease: mechanisms and assessments. Progress in Pain Research and Mangement, vol 3, IASP P, Seattle, pp 27–39, 28

[29] Borg G, Borg E (1994) Principles and experiments in category-ratio scaling. In: Reports from the Department of Psychology, Stockholm University, Sweden, pp 1–30

[30] Borg G, Borg P (1987) On the relations between category scales and ratio scales and a method for scale transformation. In: Reports from the Department of Psychology, Stockholm University, Sweden, pp 1–14, 7

[31] Borg I, Staufenbiel T (1993) Theorien und Methoden der Skalierung. (engl.: Theories and methods of scaling) Huber, Bern, p 214, 218

[32] Bosshart L (1976) Untersuchungen zur Verstehbarkeit von Radio- und Fernsehsendungen. (engl.: Experiments on comprehensibility of radio and television broadcast) In: Rundfunk und Fernsehen 24, 1–2, pp 197–209, 197

[33] Brand T, Kollmeier B (2002) Efficient adaptive procedures for threshold and concurrent slope estimates for psychophysics and speech intelligibility tests. In: J Acoust Soc Am, 111, pp 2801–2810

[34] Bronkhorst A W (2000) The cocktail party phenomenon: A review of research on speech intelligibility in multiple-talker conditions. In: acta acustica, 86, pp 117–128

[35] Brainerd C J (1978) Entwicklungsstufe, Struktur und Entwicklungstheorie. (engl.: Developmental stage, structure and theory of development) In: Steiner G (ed) Die Psychologie des 20. Jahrhunderts. Bd VII: Piaget und die Folgen. (engl.: The psychology of the 20th centry. Vol. VII: Piaget and the consequences) Kindler, Zürich, pp 207–218

[36] Bronwen L J, McManus P R (1986) Graphic scaling of qualitative terms. In: Society of motion picture and television engineers SMPTE Journal, vol 95, pp 1166–1171, 1171

[37] Burkhardt F (2001) Simulation emotionaler Sprechweise mit Sprachsynthesesystemen. (engl.: Simulation of emotional speech by means of speech synthesis systems) Shaker, Aachen

[38] Campbell W N (1999) Where is the information in speech? (and to what extent can it be modeled in synthesis?). In: Proc 3rd ESCA/COCOSDA Workshop on Speech Synthesis, Blue Mountain, Australia, CD-ROM

[39] Cherry C (1957) On human communication. A review, a survey, and a criticism. MIT P, Cambridge, Mass, p 63, 222

[40] Childers D G, Wu K, Hicks D M, Yegnanarayana B (1989) Voice conversion. In: Speech Com 8, pp 147–158

[41] Chilla R, Gabriel P, Kozielski P, Bänsch D, Kabas M (1976) Der Göttinger Kindersprachverständnistest. (engl.: Göttingen speech comprehension test for children) In: HNO 24, pp 342–346

[42] Dandi/Hcrc/Elsnet Evaluation Workshop (1992) Edinburgh: HCRC Publications, Univ of Edinburgh, Record

[43] Delogu C, Paoloni P, Pocci P, Sementina C (1991) Quality evaluation of text-to-speech synthesizers using magnitude estimation, categorical estimation, pair comparison and reaction time methods. In: Proc Eurospeech, Genoa, Italy, pp 353–355

[44] Delogu C, Paoloni P, Sementina C (1991) Comprehension of natural and synthetic speech: preliminary studies. (Final report, ESPRIT Project No. 2587, II.c)

[45] Delogu C, Conte S, Sementina C (1997) Cognitive factors in the evaluation of synthetic speech. In: Speech Com 24, pp 153–168

[46] DIN 1319 (1995–01) Grundlagen der Messtechnik. Part 1: Grundbegriffe. (engl.: Fundamentals of metrology. Part 1: Basic terms) Beuth, Berlin

[47] DIN 45621 Part 1 & 2 (July 1989) Sprache für Gehörprüfung. (engl.: Speech for testing the auditory organ) Beuth, Berlin

[48] DIN 45626 (August 1976) Tonträger zum Prüfen des Hörvermögens besprochen mit Wörtern nach DIN 45626, Aufnahme 1969. (engl.: Acoustic material for testing auditory competence, consisting of words according to DIN 45626, recording 1969) Beuth, Berlin

[49] DIN 55350 Part 11 (1995–08) Begriffe der Qualitätssicherung und Statistik. Begriffe des Qualitätsmanagements. (engl.: Terms of quality control and statistics. Terms of quality management) Beuth, Berlin

[50] DIN 55350 Part 12 (1989–03) Begriffe der Qualitätssicherung und Statistik. Merkmalsbezogene Begriffe. (engl.: Terms of quality control and statistics. Terms related to features) Beuth, Berlin

[51] DIN 55350, Part 13 (1987–07) Begriffe der Qualitätssicherung und Statistik. Begriffe zur Genauigkeit von Ermittlungsverfahren und Ermittlungsergebnissen. (engl.: Terms of quality control and statistics. Terms related to precision of investigation methods and investigative results) Beuth, Berlin

[52] DIN EN ISO 9000 (2000–12) Normen zum Qualitätsmanagement. (engl.: Standards related to quality management) Beuth, Berlin

[53] Döring W H, Hamacher V (1992) Neue Sprachverständlichkeitstests in der Klinik: Aachener Logatomtest und 'Dreisilbertest' mit Störschall. (engl.: New speech intelligibility tests in the hospital: Aachen logatom test and three syllable test with ambient noise) In: Kollmeier B (ed) Moderne Verfahren der Sprachaudiometrie. (engl.: Modern procedures of speech audiometry) Median, Heidelberg, pp 137–168

[54] Dudley H, Tarnoczy T H (1950) The speaking machine of Wolfgang von Kempelen. In: J Acoust Soc Amer 22, pp 151–166

[55] Dudley H, Riesz R R, Watkins S S A (1939) A synthetic speaker. In: J Franklin Institute, vol 227, 6, pp 739–764

[56] Dutoit T (2002) An introduction to text-to-speech synthesis. Springer, Berlin Heidelberg New York

[57] Edwards A D N (1991) Speech synthesis. Technology for disabled people. Brookes Publ, Baltimore, p 32

[58] Eisler H (1965) On psychophysics in general and the psychophysical differential equation in particular. In: Scandin J of Psych, 6, pp 85–102

[59] Eisler H (1982) On the nature of subjective scales. In: Scandin J of Psych, 6, pp 161–171

[60] Eisler H, Guirao M (1997) Conventional magnitude estimation versus converging limits procedures: A comparison. In: Preis A, Hornowski T (eds) Fechner Day 97. Proc of the 13th Annual Meeting of the Intern Soc for Psychophysics.Wydawnictwo Poznanskie, Poznan, Poland, pp 149–154

[61] Ellermeier W, Westphal W, Heidenfelder M (1991) On the »absoluteness« of category and magnitude scales of pain. In: Perception & Psychophysics 49, pp 159-166

[62] EN 60268–16 (1998) Elektroakustische Geräte. Objektive Bewertung der Sprachverständlichkeit durch den Sprachübertragungsindex. Europäisches Komitee für Elektrotechnische Normung CENELEC. (engl.: Electroacoustic instruments. Objective assessment of speech intelligibility by the speech transmission index) Brussels, Belgium

[63] Endres W (1984) Sprachsynthese: Stand der Entwicklung und vordringliche Probleme. (engl.: Speech synthesis: State-of-the-art and major problems) In: Fortschr d Akustik DAGA, Darmstadt, pp 93–116, 114

[64] Eskenazi M (1992) Changing speech styles: Strategies in read speech and casual and careful spontaneous speech. In: Proc Int Conf on Spoken Lang Proc, ICSLP, Banff, Canada, pp 755–758

[65] Eskenazi M (1993) Trends in speaking styles research. In: Proc Eurospeech, Berlin, Germany, pp 501–509

[66] Fabian R (ed) (1990) Christian von Ehrenfels. Philosophische Schriften 4. Metaphysik. (engl.: Christian von Ehrenfels. Philosophical scripts 4. Methaphysics) Philosphia Verlag, München

[67] Fant G (1984) Phonetics and Speech Technology. In: Proc Xth Intern Congress of Phonetic Sciences, Utrecht, The Netherlands, pp 13–24, 22

[68] Fechner G T (1860) Elemente der Psychophysik. (engl.: Elements of psychophysics) Breitkopf und Härtel, Leipzig

[69] Fellbaum K (1984) Sprachverarbeitung und Sprachübertragung. (engl.: Speech processing and speech transmission) Springer, Berlin Heidelberg New York, p 235

[70] Fellbaum K, Klaus H, Sotscheck J (1994) Hörversuche zur Beurteilung der Sprachqualität von Sprachsynthesesystemen für die deutsche Sprache. (engl.: Auditory experiments for assessing speech quality of speech synthesis systems for German) In: Fortschr d Akustik DAGA, Dresden, pp 117–122

[71] Flanagan J L (1964) Speech analysis: synthesis and perception. Springer, Berlin Heidelberg New York, 1972, pp 204–210

[72] Fourcin A J, Harland G, Barry W, Hazan V (eds) (1989) Speech input and output assessment. Multilingual methods and standards. Ellis Horwood, Chichester, pp 141–159

[73] Fujisaki H, Hirose K, Ohne S, Minematsu N (1990) Influence of context and knowledge on the perception of continuous speech. In: Proc Int Conf on Spoken Lang Proc, ICSLP, Kobe, Japan, pp 417–420

[74] Furui S (1989) Digital speech processing, synthesis and recognition. M Dekker, New York

[75] Furui S (2003) Toward robust speech recognition and understanding. In: Matousek V, Mautner P (eds) Lecture notes in computer science. Springer, Berlin Heidelberg New York

[76] Gescheider G A (1988) Psychological scaling. In: Annual Review of Psych, 39, pp 169–200

[77] Gibbon D (1992) Linguistic aspects of speech material complexity. (Final report, ESPRIT Project No. 2587, Se.8)

[78] Gibbon D, Moore R, Winski R (eds) (1997) Handbook of standards and resources for spoken language systems. Mouton de Gruyter, Berlin

[79] Ginsburg H, Opper S (1969) Piaget's Theorie der geistigen Entwicklung. (Piaget's theory of cognitive development) Klett-Cotta, Stuttgart 1978, pp 32–41

[80] Goldstein M (1985) Classification of methods used for assessment of text-to-speech systems according to the demands placed on the listener. In: Speech Com 16, pp 225–244

[81] Goldstein M (1993) Handbook on the assessment of text-to-speech (TTS) systems. (Final report, ESPRIT Project No. 2587)

[82] Goldstein M, Till O (1992) Is % total error a valid measure of speech synthesizer performance at the segmental level? In: Proc Int Conf on Spoken Lang Proc, ICSLP, Banff, Canada, pp 1215–1218

[83] Goldstein M, Lindström B, Till O (1992a) Assessing global performance of speech synthesizers: Context effects when assessing naturalness of Swedish sentence pairs generated by 4 systems using 3 different assessment procedures (free number magnitude estimation, 5- and 11-point category scales). (Final report, ESPRIT Project No. 2587, IIa)

[84] Goldstein M, Lindström B, Till O (1992b) Some aspects on context and range effects when assessing naturalness of Swedish sentences generated by 4 synthesizer systems. In: Proc Int Conf on Spoken Lang Proc, ICSLP, Banff, Canada, pp 1339–1342

[85] Goodman J, Nusbaum H C (1990) The effects of syntactic and discourse variables on the segmental intelligibility of speech. In: Proc Int Conf on Spoken Lang Proc, ICSLP, Kobe, Japan, pp 393–396

[86] Grave H F (1971) Größen, Einheiten und Gleichungen. (engl.: Measurands, units and equations) In: Sonderdruck L3396, Elektro-Anzeiger, 24th vol, No. 10, pp 227–229

[87] Greenspan S L, Nusbaum H C, Pisoni D B (1985) Perception of speech generated by rule: Effects of training and attentional limitations. (Research on Speech Perception Progress Report 11, Indiana University, USA, pp 263–287)

[88] Gregory R L (1966) Eye and brain. The psychology of seeing. Weidenfeld & Nicolson, London, p 224

[89] Grice M, Hazan V (1989) The assessment of synthetic speech intelligibility using semantically unpredictable sentences. (Speech, Hearing & Language: Work in progress, UCL, vol 3, London, Great Britain, pp 109–122)

[90] Grice M, Vagges K, Hirst D (1991) Prosodic form tests. (Final report, ESPRIT Project No. 2587, III.91)

[91] Gross S (1994) Lese-Zeichen. Kognition, Medium und Materialität im Leseprozess. (engl.: Book-marks. Cognition, medium and materialism in the reading process) Wiss Buchgesellschaft, Darmstadt, p 3

[92] Guilford J P (1954) Psychometric methods. Mc Graw, New York

[93] Guski R (1996) Wahrnehmen – ein Lehrbuch. (engl.: Perceiving – a text book) Kohlhammer, Stuttgart

[94] Hagerman B (1982) Sentences for testing speech intelligibility in noise. In: Scand Audiol, 11, pp 79–87

[95] Hagerman B (1984) Clinical measurement of speech reception threshold in noise. In: Scand Audiol, 13, pp 57–63

[96] Hahlbrock K H (1953) Über Sprachaudiometrie und neue Wörterteste. (engl.: About speech audiometry and new tests using words) In: Archiv f Ohren-, Nasen- und Kehlkopfheilkunde 162, pp 394–432

[97] Hahlbrock K H (1979) Sprachaudiometrie. Grundlagen und praktische Anwendung einer Sprachaudiometrie für das deutsche Sprachgebiet. (engl.: Speech audiometry. Fundamentals and practical use of speech audiometry for the German language) Thieme, Stuttgart

[98] Halka U (1993) Objektive Qualitätsbeurteilung von Sprachkodierverfahren unter Anwendung von Sprachmodellprozessen. (engl.: Objective quality assessment of speech coding systems using speech model processes) Ph.D. thesis, Ruhr-University Bochum

[99] Halka U, Heute U (1992) A new approach to objective quality-measures based on attribute-matching. In: Speech Com 11, pp 15–30

[100] Hansen M, Kollmeier B (2000) Objective modelling of speech quality with a psychoacoustically validated auditory model. In: J Audio Eng Soc 48(5), pp 395–409

[101] Haubensak G (1985) Absolutes und vergleichendes Urteil. Eine Einführung in die Theorie psychischer Bezugssysteme. (engl.: Absolute and relative judgment. An introduction to the theory of psychological reference systems) Springer, Berlin Heidelberg New York

[102] Hauenstein M (1997) Psychoakustisch motivierte Maße zur instrumentellen Sprachgütebeurteilung. (engl.: Psycho-acoustically motivated measures for instrumental speech quality assessment) Shaker, Aachen

[103] Hazan V, Shi B (1993) Individual variability in the perception of synthetic speech. In: Proc Eurospeech, Berlin, Germany, pp 1849–1852

[104] Heinrichs R (1993) Entwicklung von Testverfahren zur Beurteilung charakteristischer akustischer Übertragungseigenschaften von Freisprecheinrichtungen. (engl.: Development of test methods for judging typical acoustic transmission characteristics of hands-free terminals) Dipl. thesis, Ruhr-University Bochum

[105] Hellbrück J, Ellermeier W (2004) Hören. Physiologie, Psychologie und Pathologie. (engl.: Hearing. Physiology, psychology and pathology) Hogrefe, Göttingen, 2nd ed.

[106] Henle M (ed) (1971) The selected papers of Wolfgang Köhler. Livernight, New York

[107] Hermann T (1988) Sprachproduktion als Systemregulation. (engl.: Speech production as system regulation) In: Blanken G, Dittmann J, Wallesch C W (eds) Sprachproduktionsmodelle. Neuro- und psycholinguistische Theorien der menschlichen Spracherzeugung. (engl.: Speech production models. Neuro- and psycho-linguistic theories of human speech production) Hochschulverlag, Freiburg, pp 19–34

[108] Heute U (1990) Sprachverarbeitung. Skriptum der Arbeitsgruppe Digitale Signalverarbeitung. (engl.: Speech processing. Script of the digital speech processing research group) Ruhr-University Bochum, Germany, pp 3–4

[109] Hörmann H (1978) Meinen und Verstehen. Grundzüge einer psychologischen Semantik. (engl.: Intenting and comprehending. Fundamentals of a psychological semantics) Suhrkamp, Frankfurt aM, p 425

[110] House A, Williams C, Hecker M, Kryter K (1965) Articulation testing methods: consonantal differentiation with a closed response set. In: J Acoust Soc Am 37, pp 158–166

[111] Howard-Jones P (1992a) SOAP. Speech output assessment package. Version 4.0. (ESPRIT Project No. 2587)

[112] Howard-Jones P (1992b) Specification of listener dimensions. (Final report, ESPRIT Project No. 2587, So. 9)

[113] ISO 8402 E/F/R (1994), International standard. Quality management and quality assurance – Vocabulary. Geneva, Switzerland, p 6

[114] ITU–T Contr COM 12–101 (1999) Performance analysis of the Call Clarity Index (CCI) and non-intrusive version of the E-model over a range of network conditions. Intern Telecom Union, Geneva, Switzerland

[115] ITU–T Contr COM 12–C1–E (2004) Questions allocated to Study Group 12 (Performance and quality of servive) for the 2005–2008 Study Period. Question 13/12 – Multimedia QoE/QoS performance requirements and assessment methods. Intern Telecom Union, Geneva, Switzerland, pp 22–23

[116] ITU–T Rec G. 107 (2003) The E-Model, a computational model for use in transmission planning. Intern Telecom Union, Geneva, Switzerland

[117] ITU–T Rec P.561 (2002) In-Service, Non-Intrusive Measurement Device – voice service measurements. Intern Telecom Union, Geneva, Switzerland

[118] ITU–T Rec P.562 (2004) Analysis and interpretation of INMD voice-services measurements. Intern Telecom Union, Geneva, Switzerland

[119] ITU–T Rec P.800 (1996), Telephone Transmission Quality. Methods for objective and subjective assessment of quality. Intern Telecom Union, Geneva, Switzerland

[120] ITU–T Rec P.861 (1996) Objective quality measurement of telephone-band (300–3400 Hz) speech codecs. Intern Telecom Union, Geneva, Switzerland

[121] ITU–T Rec P.862 (2001) Perceptual evaluation of speech quality (PESQ), an objective method for end-to-end speech quality assessment of narrowband telephone networks and speech codecs. Intern Telecom Union, Geneva, Switzerland

[122] ITU–T Suppl 3 to P.-Series Rec (1993) Models for predicting transmission quality from objective measurements. Intern Telecom Union, Geneva, Switzerland

[123] Jakobson R (1969) Kindersprache, Aphasie und allgemeine Lautgesetze. (engl.: Child language, aphasia and language universals) Suhrkamp, Frankfurt aM

[124] Jakobson R, Waugh L (1979) The sound shape of language. Harvester P, Brighton, p 234

[125] Jekosch U (1989) German Study. (Final report, ESPRIT Project No. 2587, 77–82)

[126] Jekosch U (1992a) The Cluster-Identification Test. (Final report, ESPRIT Project No. 2587, 1.III.91–28.II.1992)

[127] Jekosch U (1992b) The Cluster-Identification Test. In: Proc Int Conf on Spoken Lang Proc, ICSLP, Banff, Canada, pp 205–209

[128] Jekosch U (1993) Speech quality assessment and evaluation. In: Proc Eurospeech, Berlin, Germany, pp 1387–1394

[129] Jekosch U (1994a) Speech intelligibility testing: On the interpretation of results. In: J of the Am Voice I / O Society, 15, pp 63–79

[130] Jekosch U (1994b) Sprachverständlichkeit in Relation zum erwartungsbezogenen Schätzwert. (engl.: Speech intelligibility in relation to an expectation-related estimation value) In: Fortschr d Akustik DAGA, Dresden, pp 1357–1360

[131] Jekosch U (1997) A structured way of looking at the performance of text-to-speech systems. In: Van Santen J P H, Sproat R W, Olive J P, Hirschberg J (eds) Progress in speech synthesis. Springer, Berlin Heidelberg New York, pp 519–527

[132] Jekosch U (2000) Sprachqualitätsbeurteilung im Anwendungszusammenhang. (engl.: Speech quality judgment in application contexts) In: Fortschr d Akustik DAGA, Oldenburg, pp 202–203

[133] Jekosch U (2005) Assigning meaning to sounds: semiotics in the context of product-sound-design. In: Blauert J (ed) Communication acoustics. Springer, Berlin Heidelberg New York

[134] Jekosch U, Belhoula A (1991) Rechnergestützte Generierung von Testmaterialien zur Sprachgütebeurteilung. (engl.: Computer-based generation of test vocabularies for speech quality judgments) In: Fortschr d Akustik DAGA Bochum, pp 905–908

[135] Jekosch U, Bodden M (1997) Verification of the E-model: results of a pilot study. Intern Telecom Union, Study Group 12 – Contr 13, Geneva, Switzerland

[136] Jekosch U, Pols L C W (1994) A feature-profile for application-specific speech synthesis assessment and evaluation. In: Proc Int Conf on Spoken Lang Proc, ICSLP, Kobe, Japan, pp 1319–1322

[137] Jo C W, Kim K T, Lee Y J (1994) Generation of multi-syllable nonsense words for the assessment of Korean text-to-speech system. In: Proc Int Conf on Spoken Lang Proc, ICSLP, Yokohama, Japan, pp 1255–1258

[138] Jones B J, Marks L E (1985) Picture quality assessment: A comparison of ratio and ordinal scales. In: SMPTE J, pp 1244–1248

[139] Jongenburger W, Van Bezooijen R (1992) Evaluatie van ELK: attitudes van de gebruikers, verstaanbaarheid en acceptabiliteit van de spraaksynthese, bruikbaarheid van het zoeksysteem. (engl.: Evaluation of ELK: user expectations, intelligibility and acceptability of the speech synthesis, usability of the search system) Stichting Spraaktechnologie, Utrecht

[140] Kalikow D N, Steven K N, Elliot L L (1977) Development of a test of speech intelligibility in noise using sentences with controlled word predictability. In: J Acoust Soc Am, 61, pp 1337–1351

[141] Kasuya H, Kasuya S (1992) Relationships between syllable, word and sentence intelligibilities of synthetic speech. In: Proc Int Conf on Spoken Lang Proc, ICSLP, Banff, Canada, pp 1215–1218

[142] Katz D (1948) Gestaltpsychologie. (engl.: Gestalt psychology) Schwabe, Stuttgart, pp 33–39

[143] Kettler F, Hottenbacher A, Heinrichs R, Gierlich H W (1994) Verfahren zur subjektiven Qualitätsbeurteilung von Freisprecheinrichtungen. (engl.: Methods for subjective quality judgment of hands-free terminals) In: Fortschr d Akustik DAGA, Dresden, pp 1365–1368

[144] Kliem K (1990) Testverfahren zur Erfassung der sprachlichen Hörfähigkeit und ihrer Langzeitveränderung bei Hörgeschädigten. (engl.: Test methods for monitoring speech perception ability and its long-term change of hearing-impaired people) MA thesis, Christian-Albrechts University Kiel

[145] Kliem K, Kollmeier B (1992) Ein Zweisilber-Reimtest in deutscher Sprache. (engl.: A two-syllable rhyme test for German) In: Kollmeier B (ed) Moderne Verfahren der Sprachaudiometrie. (engl.: Modern procedures of speech audiometry) Median, Heidelberg, pp 287–310

[146] Koffka K (1935) Principles of Gestalt psychology. Lund Humphries, London

[147] Kohler K J (1986) Invariance and variability in speech timing: From utterance to segment in German. In: Perkell J S, Klatt D H (eds) Invariance and variability in speech processes. Erlbaum, Hillsdale, pp 268–289

[148] Kohonen T (1989) Self-organizing and associative memory. Springer, Berlin Heidelberg New York, pp 1–29

[149] Kollmeier B (1990) Messmethodik, Modellierung und Verbesserung der Verständlichkeit von Sprache. (engl.: Measuring method, modeling and improvement of speech intelligibility) Habil thesis, Georg-August-University Göttingen

[150] Kollmeier B, Wesselkamp M (1997) Development and evaluation of a German sentence test for objective and subjective speech intelligibility assessment. In: J Acoust Soc Am, 102, pp 2412–2421

[151] Koopmans-Van Beinum F (1992) The role of focus words in natural and synthetic continuous speech: acoustic aspects. In: Speech Com 11, pp 439–452, 440

[152] Kornatzki von P (1989) Text und Bild. (engl.: Text and picture) In: Stankowski A, Duschek K (eds) Visuelle Kommunikation. (engl.: Visual communication) Reimer, Berlin, pp 177–208

[153] Köster S, Pörschmann C, Walter J (2000) Eine Datenbank für deutsche Sprache mit Lombard-Effekt. (engl.: A database for the German language with Lombard-effect) In: Fortschr d Akustik DAGA, Oldenburg, pp 356–357

[154] Kraft V, Portele T (1995) Quality evaluation of five speech synthesis systems for German. In: acta acustica, 3, pp 351–366

[155] Kratzenstein C G (1782) Sur la naissance et la formation des voyelles. (engl.: On the origin and the formation of vowels) In: J de Physique 21, pp 358–380

[156] Kröber G (1968) Die Kategorie 'Struktur' und der kategorische Strukturalismus. (engl.: The category 'structure' and the categorical structuralism) In: Dtsch Z f Philosophie 16, pp 1310–25, 1314

[157] Lazarus H (1990) New methods for describing and assessing direct speech communication under disturbing conditions. In: Environmental Intern, 16, pp 373–392

[158] Lenke N, Lutz H D, Sprenger M (1995) Grundlagen sprachlicher Kommunikation. Mensch. Welt. Handeln. Sprache. Computer. (engl.: Fundamentals of speech communication. Man. World. Acting. Speech. Computer) Fink, München

[159] Leontjew, A A (1971) Sprache – Sprechen – Sprechtätigkeit. (engl.: Language – speaking – acting with speech) Kohlhammer, Stuttgart, p 31

[160] Lewandowski T (1980) Linguistisches Wörterbuch. (engl.: Linguistic dictionary) Quelle & Meyer, Heidelberg, p 968, 994

[161] Lewis E, Tatham M (1999) Word and syllable concatenation in text-to-speech synthesis. In: Proc Eurospeech, Budapest, Hungary, pp 615–618

[162] Lieberman P (1963) Some effects of semantic and grammatical context on the production and perception of speech. In: Language and Speech, 6, pp 172–187

[163] Lienert G A, Raatz U (1961) Testaufbau und Testanalyse. (engl.: Test design and test analysis) Beltz, Weinheim

[164] Logan J, Greene B, Pisoni D B (1989) Segmental intelligibility of synthetic speech produced by rule. In: J Acoust Soc Am, 86(2), pp 566–581

[165] Luchins A S, Luchins E H (1982) An introduction to the origins of Wertheimer's Gestalt psychology. In: Gestalt Theory, 4 (3–4), pp 145–171

[166] Luce R D, Krumhansl C L (1988) Measurement, scaling and psychophysics. In: Atkinson R C, Herrnstein R J, Lindzey G, Luce R D (eds) Steven's Handbook of Experimental Psychology, vol 1: Perception and Motivation. Wiley & Sons, New York, pp 3–74

[167] Luce P A, Feustel P C, Pisoni D B (1983) Capacity demands in short-term memory for synthetic and natural speech. In: Human Factors, 25, pp 17–32

[168] Lucid D P (1977) Introduction. In: Lucid D P (ed) Soviet Semiotics, Johns Hopkins Univ P, Baltimore, pp 1–23, 7

[169] Lüschow F (1992) Sprache und Kommunikation in der technischen Arbeit. (engl.: Language and communication in the technical working context) Lang, Frankfurt, p 6

[170] Malter B (1995) Erstellung eines Programmsystems zur Auswertung der CLID-Testergebnisse. (engl.: Design of a computer system to analyze CLID-test results) Dipl. thesis, Ruhr-University Bochum

[171] Marks L E, Borg G, Ljunggren G (1983) Individual differences in perceived exertion assessed by two new methods. In: Perception & Psychophysics, 34 (3), pp 280–288

[172] Massaro D W, Cohen M M (1990) The joint influence of stimulus information and context in speech perception. In: Proc Int Conf on Spoken Lang Proc, ICSLP, Kobe, Japan, pp 413–416

[173] Matlin M W, Foley H J (1983) Sensation and perception. Allyn & Bacon, Boston

[174] Merleau-Ponty M (1966) Phänomenologie der Wahrnehmung. (engl.: Phenomenology of perception) de Gruyter & Co, Berlin, pp 22–23

[175] Mersdorf J (1995) Untersuchungen zur perzeptiven Unterscheidbarkeit der Prosodie verschiedener Sprechereignisse. (engl.: Analyses of the perceptual distinctness of prosody of different speech events) In: Fortschr d Akustik DAGA, Saarbrücken, pp 1007–1010

[176] Middelweerd M J, Fetsen J M, Plomp R (1990) Difficulties with speech intelligibility in noise in spite of normal pure-tone audiogram. In: Audiology 29, pp 1–7

[177] Möller S (1997a) Development of scenarios for a short conversation test. Intern Telecom Union, Study Group 12 – Contr 35

[178] Möller S (1997b) The E-Model: An analysis of the sources and comparison with published and new test results. Intern Telecom Union, Study Group 12 – Contr 37

[179] Möller S (1999) Assessment and prediction of telephone speech quality. Ph.D. thesis, Ruhr-University Bochum

[180] Möller S (2000) Assessment and prediction of speech quality in telecommunications. Kluwer Academic Publ., Boston

[181] Möller S (2005) Quality of transmitted speech for humans and machines. In: Blauert J (ed) Communication acoustics. Springer, Berlin Heidelberg New York

[182] Möller S (2005) Quality of telephone-based spoken dialogue systems. Springer, Berlin Heidelberg New York

[183] Möller S, Raake A (2000) Planung von Telefon-Sprachqualität: Vorstellung von unterschiedlichen Modellansätzen. (engl.: Planning of telephone speech quality: introduction of different model approaches) In: Fortschr d Akustik DAGA, Oldenburg, pp 208–209

[184] Moles A A (1991) Design und Immaterialität. Was hat es damit in einer postindustriellen Gesellschaft auf sich? (engl.: Design and immateriality. How is it dealt with in the post-industrial society?) In: Rötzer F (ed) Digitaler Schein. Ästhetik der elektronischen Medien. (engl.: Digital illusion. Aesthetics of electronic media) Suhrkamp, Frankfurt aM, pp 160–170, 169

[185] Morimoto T, Shikano K, Iida H, Kurematsu A (1990) Integration of speech recognition and language processing in the spoken language translation system SL-TRANS. In: Proc Int Conf on Spoken Lang Proc, ICSLP, Kobe, Japan, pp 921–928

[186] Ney H (2003) Maschinelle Sprachverarbeitung: Der statistische Ansatz in der Spracherkennung und Sprachübersetzung. (engl.: Automatic speech processing: the statistic approach in speech recognition and translation) In: Informatik Spektrum, Vol. 26, 2, pp 94–102

[187] Nikléczy O, Olaszy G (2003) A reconstruction of Farkas Kempelen's speaking machine. In: Proc Eurospeech, Geneva, Switzerland, pp 2453–2456

[188] Nilsson M, Soli S D, Sullivan J A (1994) Development of Hearing In Noise Test for measurement of speech reception thresholds in quiet and in noise. In: J Acoust Soc Am, 95, pp 1085–1099

[189] Nöth W (1977) Dynamik semiotischer Systeme. Vom altenglischen Zauberspruch zum illustrierten Werbetext. (engl.: Dynamics of semiotic systems. From the old English charm to illustrated texts in advertisements) Metzler, Stuttgart, p 83

[190] Nöth W (2000) Handbuch der Semiotik. (engl.: Handbook of semiotics) Metzler, Stuttgart, 227

[191] Nusbaum H C (1990) The role of learning and attention in speech perception. In: Proc Int Conf on Spoken Lang Proc, ICSLP, Kobe, Japan, pp 409–412

[192] Nye P, Gaitenby J (1974) The intelligibility of synthetic monosyllabic words in short, syntactically normal sentences. (Status Report on Speech Research, 37/38, pp 169–190 by Haskins Laboratories)

[193] Ozawa K, Logan J S (1987) The perception of digitally coded speech by native and non-native speakers of English. In: Res on Speech Perception Progress, 13, pp 71–100

[194] Pavlovic C V (1987) Derivation of primary parameters and procedures for use in speech intelligibility predictions. In: J Acoust Soc Am 82, No. 2, pp 413–433

[195] Pavlovic C, Rossi M, Espesser R (1989) A comparative analysis of the magnitude estimation and pair comparison techniques for use in assessing quality of text-to-speech synthesis. In: Proc ESCA-Workshop on Speech I/O Assessment and Speech Databases, Noordwijkerhout, The Netherlands, pp 3.2–3.6

[196] Pavlovic C, Rossi M, Espesser R (1990) Use of the magnitude estimation technique for assessing the performance of text-speech synthesis systems. In: J Acoust Soc Am 87, pp 373–382

[197] Pavlovic C, Sorin C, Roumiguiere J P, Lucas J P (1990) Cross validation between a magnitude estimation technique and a pair comparison technique for assessing quality of text-to-speech synthesis systems. In: J d'Acoustique 3, pp 75–83

[198] Peckels J, Rossi M (1973) Le test diagnostic par paires minimales. Adaptation au Francais du 'Diagnostic Rhyme Test' de W. D. Voiers. (engl.: The diagnostic rhyme test using minimal pairs. Adaption of the diagnostic rhyme test of W. D. Voiers to French language) In: Revue d'Acoustique 27, pp 245–262

[199] Pfister B (2001) Personenidentifizierung anhand der Stimme. (engl.: Person identification on the basis of voice) In: Kriminalistik 55, 4, pp 287–291

[200] Pfister B, Beutler R (2003) Estimating the weight of evidence in forensic speaker verification. In: Proc Eurospeech, Geneva, Switzerland, pp 701–704

[201] Pfister B, Traber C (1994) Text-to-speech synthesis: an introduction and a case study. In: Keller E (ed) (1994) Fundamentals of speech synthesis and speech recognition. Basic concepts, state of the art and future challenges. Wiley Sons, Chichester, pp 87–107, 105

[202] Piaget J (1924) Le jugement et le raisonnement chez l'enfant. (engl.: Judgment and reasoning in the child) Routledge & Kegan, London, 1962

[203] Piaget J (1937) La construction réel chez l'enfant. (engl.: The child's construction of reality) Klett, Stuttgart, 1974

[204] Piaget J (1972) The principles of genetic epistemology. Basic B, New York

[205] Pijper J R de (1997) High-quality message-to-speech generation in a practical application. In: Van Santen J P H, Sproat R W, Olive J P, Hirschberg J (eds) Progress in speech synthesis. Springer, Berlin Heidelberg New York, pp 575–586

[206] Pisoni D B, Garber E E (1990) Lexical memory in visual and auditory modalities: The case for a common mental lexicon. In: Proc Int Conf on Spoken Lang Proc, ICSLP, Kobe, Japan, pp 401–405

[207] Pisoni D B, Manous L M, Dedina M J (1987) Comprehension of natural and synthetic speech: Effects of predictability on the verification of sentences controlled for intelligibility. In: Computer, Speech and Language, 2, pp 303–320

[208] Pisoni D B, Remez R (2004) The handbook of speech perception. Blackwell, Oxford

[209] Plomp R, Mimpen A M (1979) Improving the reliability of testing the speech perception threshold for sentences. In: Audiology, 8, pp 43–52

[210] Pörschmann C (2000) Einfluss der Eigenwahrnehmung der Stimme auf die Präsenz in auditiven virtuellen Umgebungen. (engl.: The influence of the perception of one's own voice on presence in auditory virtual environments) In: Fortschr d Akustik DAGA, Oldenburg, pp 164–165

[211] Price D D (1988) Psychological and neural mechanisms of pain. Raven P, New York

[212] Quackenbush T P, Barnwell T P, Clements M A (1988) Objective measures of speech quality. Prentice Hall, New Jersey

[213] Raake A (2004) Predicting speech quality under random packet loss: Individual impairment and additivity with other network impairments. In: Jekosch U, Möller S (eds) Special issue on auditory quality of systems, acta acustica united with Acustica 90 (6), pp 1061–1083

[214] Raake A (2005) Assessment of parametric modelling of speech quality in voice-over-IP networks. Ph.D. thesis, Ruhr-University Bochum

[215] Reitzle W (1993) Qualität – Wichtige Komponente der F- und E-Strategie eines Automobilherstellers. (engl.: Quality – An important component of the R&D-strategy of a car manufacturer) In: Seghezzi H D, Hansen J R (eds) Qualitätsstrategien. Anforderungen an das Management der Zukunft. (engl.: Quality strategies. Requirements to the management of future) Hanser V, München, pp 94–108, 94

[216] Richards D L (1973) Telecommunication by speech. Butterworth & Co Ltd, London

[217] Rietveld A C M, Lankhorst E, Speyer R (1989) Subjects as a factor in the identification of synthetic speech stimuli. (AFN Proc Univ of Nijmegen, 13, pp 25–30)

[218] Rinscheid A (1996) Voice conversion based on topological feature maps and time-variant filtering. In: Proc Int Conf on Spoken Lang Proc, ICSLP, Philadelphia, USA, pp 1445–1448

[219] Rossi M, Espesser R, Pavlovic C (1991) The effects of internal reference system and cross-modality matching on the subjective rating of speech synthesizers. In: Proc Eurospeech, Genoa, Italy, pp 273–276

[220] Rosson M B (1985) Listener training for speech-output applications. In: Proc Conf on Human Factors in Computing Systems, pp 193–196

[221] Sagisaka Y, Campbell N, Higuchi N, Sagisaga Y (eds) (1997) Computing prosody. Computational models for processing spontaneous speech. Springer, Berlin Heidelberg New York

[222] Sahrhage J, Blauert J (2000) Interaktive auditive virtuelle Umgebungen. (engl.: Interactive auditory virtual environments) In: Fortschr d Akustik DAGA, Oldenburg, pp 150–151

[223] Saussure F de (1916) Cours de linguistique générale. (engl.: Course in general linguistics) de Gruyter, Berlin, 1967, 79

[224] Schröder M R (1993) A brief history of synthetic speech. In: Speech Com 13, pp 231–237

[225] Schröter J, Cosatto E, Graf H P, Ostermann J (2004) From audio-only to audio and video text-to-speech. In: Jekosch U, Möller S (eds) Special issue on auditory quality of systems, acta acustica united with Acustica 90 (6), pp 1084–1095

[226] Schubert K (1958) Sprachhörprüfmethoden. Grundlagen. (engl.: Assessment methods for the speech hearing capacity. Fundamentals) Thieme, Stuttgart

[227] Schulte-Fortkamp B (1994) Geräusche beurteilen im Labor. Entwicklung interdisziplinärer Forschungsmethoden und ihre forschungssoziologische Analyse. (engl.: Judging sounds in laboratory environments. Development of interdisciplinary research methods and their sociological analysis) VDI-Verlag, Düsseldorf

[228] Schwab E C, Nusbaum H C, Pisoni D B (1985) Effects of training on the perception of synthetic speech. In: Human Factors 27, pp 395–408

[229] Seiler T B (1978) Grundlegende Entwicklungstätigkeiten und ihre regulative, systemerzeugende Interaktion. (engl.: Basic developmental actions and their regulating system-generating interaction) In: Steiner G (ed) Die Psychologie des 20. Jahrhunderts. Bd. VII: Piaget und die Folgen, (engl.: The psychology of the 20th century. Vol. VII: Piaget and the consequences) Kindler, Zürich, pp 628–645, 629

[230] Silverman K, Basson S, Levas S (1990) Evaluating synthesizer performance: Is segmental intelligibility enough? In: Proc Int Conf on Spoken Lang Proc, ICSLP, Kobe, Japan, pp 981–984

[231] Simon C, Fourcin A J (1976) Differences between individual listeners in their comprehension of speech and perception of sound patterns. In: Speech, Hearing and Language: Work in progress, UCL, vol 2, London, Great Britain, pp 94–125

[232] Sixtl F (1967) Messmethoden der Psychologie. (engl.: Measuring methods in psychology) Beltz, Weinheim

[233] Smith B (ed) (1988) Foundations of Gestalt Theory. Philosophia V, Vienna

[234] Sonntag G P (2000) Evaluation von Prosodie. (engl.: Evaluation of prosody) Shaker, Aachen

[235] Sonntag G P, Portele T (1998) PURR – a method for prosody evaluation and investigation. In: J ComputerSpeech and Language, vol 12, No. 4, pp 437–451

[236] Sotscheck J (1982) Ein Reimtest für Verständlichkeitsmessungen mit deutscher Sprache als ein verbessertes Verfahren zur Bestimmung der Sprachübertragungsgüte. (engl.: A rhyme test for intelligibility measurements for German as an improved method for the determination of speech transmission quality) In: Der Fernmeldeingenieur 36, pp 1–84

[237] Sotscheck J (1989) Sprachverständlichkeit bei additiven Störungen. (engl.: Speech intelligibility under additive noise) In: Acustica 57, 4/5, pp 257–267

[238] Spiegel M, Altom M, Macchi M, Wallace K (1990) Comprehensive assessment of the telephone intelligibility of synthesized and natural speech. In: Speech Com 9, pp 279–291

[239] Stankowski A (1989) Visualisierung. (engl.: Visualization) In: Stankowski A, Duschek K (eds) Visuelle Kommunikation. (engl.: Visual communication) Reimer, Berlin, pp 19–52, 50–51

[240] Steeneken H J M, Houtgast T (1980) A physical method for measuring speech transmission quality. In: J Acoust Soc Am 67, pp 318–326

[241] Stevens S S, Galanter E (1957) Ratio scales and category scales for a dozen perceptual continua. In: J of Experim Psych, 16, pp 377–411

[242] Stevens S S (1988) Handbook of experimental psychology. Wiley & Sons, New York

[243] Stöber K, Portele T, Wagner P, Hess W (1999) Synthesis by word concatenation. In: Proc Eurospeech, Budapest, Hungary, pp 619–622

[244] Styger T, Keller E (1994) Formant Synthesis. In: Keller E (ed) (1994) Fundamentals of speech synthesis and speech recognition. Basic concepts, state of the art and future challenges. Wiley & Sons, Chichester, pp 109–128, 113

[245] Tatham M, Lewis E, Morton K (1999) An advanced intonation model for synthesis. In: Proc Eurospeech, Budapest, Hungary, pp 1871–1874

[246] Trautmann U, Langhoff T (1994) Die Beeinträchtigung der direkten sprachlichen Kommunikation – Ein Belastungsfaktor im Arbeitsprozess. (engl.: The impairment of direct speech communication – a straining factor in working processes) In: Fortschr d Akustik DAGA, Dresden, pp 1457–1460

[247] Tschopp K, Ingold L (1992) Die Entwicklung der deutschen Version des SPIN-Tests (Speech Perception in Noise). (engl.: The SPIN test: development of the German version) In: Kollmeier B (ed) Moderne Verfahren der Sprachaudiometrie. (engl.: Modern procedures of speech audiometry) Median, Heidelberg, pp 311–329

[248] Van Bezooijen R (1988) Evaluation of two synthesis systems for Dutch – intelligibility and overall quality of initial and final consonant clusters. (SPIN-ASSP report No. 5, Utrecht: Stichting Spraaktechnologie)

[249] Van Bezooijen R, Van Heuven V J (1997) Assessment of speech output systems. In: Gibbon D, Moore R, Winski R (eds) Handbook of standards and resources for spoken language systems. De Gruyter, Berlin, pp 481–563

[250] Van Erp A, Grice M (1989) Multi-lingual syntactic structures for semantically unpredictable sentences. (Final report, ESPRIT Project No. 2587, pp 43–60)

[251] Vary P, Heute U, Hess W (1998) Digitale Sprachsignalverarbeitung. (engl.: Digital speech signal processing) Teubner, Stuttgart, pp 528–536

[252] Voiers W (1983) Evaluating processed speech using the diagnostic rhyme test. In: Speech Tech 1, pp 338–352

[253] Vollmer G (1975) Evolutionäre Erkenntnistheorie. Angeborene Erkenntnis-strukturen im Kontext von Biologie, Psychologie, Linguistik, Philosophie und Wissenschaftstheorie. (engl.: Evolutionary epistemology. Inborn epistemic structures in the context of biology, psychology, linguistics, philosophy and theory of sciences) Hirzel, Stuttgart, p 16, 37, 125, 180

[254] Weizsäcker von E (ed) (1974) Offene Systeme I. (engl.: Open systems I) Klett, Stuttgart, p 10

[255] Wagener K, Brand T, Kollmeier B (1999a) Entwicklung und Evaluation eines Satztests für die deutsche Sprache I: Design des Oldenburger Satztests. (engl.: Development and evaluation of a sentence test for German I: Design of the Oldenburg sentence test) In: Zeitschr f Audiol, 38, pp 4–15

[256] Wagener K, Brand T, Kollmeier B (1999b) Entwicklung und Evaluation eines Satztests für die deutsche Sprache II: Optimierung des Oldenburger Satztests. (engl.: Development and evaluation of a sentence test for German II: An optimization of the Oldenburg sentence test) In: Zeitschr f Audiol, 38, pp 44–56

[257] Wagener K, Brand T, Kollmeier B (1999c) Entwicklung und Evaluation eines Satztests für die deutsche Sprache III: Evaluation des Oldenburger Satztests. (engl.: Development and evaluation of a sentence test for German III: Evaluation of the Oldenburg sentence test) In: Zeitschr f Audiol, 38, pp 86–95

[258] Wagner P S, Breuer S, Stöber K (2000) Automatische Prominenzetikettierung einer Datenbank für die korpusbasierte Sprachsynthese. (engl.: Automatic labeling of prominence of a database for data-driven speech synthesis) In: Fortschr d Akustik DAGA, Oldenburg, pp 350–351

[259] Wahlster W (1993) Verbmobil, translation of face-to-face dialogs. In: Proc Eurospeech, Berlin, Germany, pp 29–38

[260] Waibel A, Jain A N, Mc Nair A E, Saito H, Hauptmann A, Tebelski J (1991) A speech-to-speech translation system using connectionist and symbolic processing strategies. In: Proc IEEE Intern Conf on Acoustics, Speech, and Signal Proc, ICASSP, pp 793–796

[261] Wang S, Sekey A, Gersho A (1992) An objective measure for predicting subjective quality of speech coders. In: IEEE J on Selected Areas in Communication, 10(5), pp 819–829

[262] Ward L M (1991) Associative measurement of psychological magnitude. In: Bolanowski S J, Gescheider G A (eds) Ratio scaling of psychological magnitude: In honour of the memory of S. S. Stevens. Erlbaum, Hillsdale, pp 79–100

[263] Warren H C (1934) Dictionary of psychology. The Riverside P, Boston, p 275

[264] Wesselkamp M (1994) Messung und Modellierung der Verständlichkeit von Sprache. (engl.: Measurement and modeling of speech intelligibility) Ph.D. thesis, University Göttingen

[265] Zimbardo P G (1988) Psychologie. (engl.: Psychology) Springer, Berlin Heidelberg New York, p 286

[266] Zwislocki J J (1983) Absolute and other scales: Question of validity. In: Perception and Psychophysics, 33, pp 593–594

[267] Zwislocki J J (1997) On some established and potential laws of auditory psychophysics. In: Preis A, Hornowski T (eds) Fechner Day 97. Proc of the 13th Annual Meeting of the Intern Soc for Psychophysics.Wydawnictwo Poznanskie, Poznan, Poland, pp 67–72, 93

[268] Zwislocki J J, Goodman D A (1980) Absolute scaling and sensory magnitudes: a validation. In: Perception and Psychophysics, 28, pp 28–38

Definitions

»controlled perception«
Result of reflection on an externally initiated (extrinsic) perception event that has been artificially evoked. (cf. page 14)

»controlled quality assessment«
The extrinsically initiated process of comparing the entirety of the perceived features of an observed, representative entity in terms of its suitability to fulfil all the individual's and/or group's expectations of features. The result of this comparative process is the quality assessment. (cf. page 14)

»desired composition«
Totality of features of individual expectations and/or relevant demands and/or social requirements. (cf. page 16)

»feature«
Recognizable and nameable characteristic of an entity. (cf. page 16)

»measurand«
Feature of an object to be measured which can numerically be described in the course of the measuring process. (cf. page 61)

»measuring«
The entirety of all the activities in the measurement chain up to determining the value of a dimension. (cf. page 63)

»measuring process«
A measuring process is an observing, assessing or instrumental process, a calculating, statistical estimating process, or a combination of the above. What is determined is an assessment, observation, an instrumental measurement, calculation or a combination of these. Depending on which kind of measuring process is used, the result of the measurement is termed an assessed, observed, instrumentally registered, calculated or statistically estimated result. (cf. page 64)

»perceived composition«
Totality of features of an entity. Signal for the identity of the entity visible to the perceiver. (cf. page 16)

»quality«
Result of judgment of the perceived composition of an entity with respect to its desired composition. (cf. page 15)

»random perception«
Result of reflection on a casual (intrinsic) perception event that has been evoked naturally. (cf. page 14)

»random quality assessment«

The intrinsically initiated process of comparing the totality of the perceived features of an observed, representative entity in terms of its suitability to fulfil all the individual's or group's expectations of features. The result of this comparative process is the quality assessment.   (cf. page 14)

»scaling«

The entirety of all the activities that are concretely applied to the process of assigning a value to a dimension, which vehicle is the measuring object, on a corresponding scale value (measuring value) according to set rules.   (cf. page 63)

»speech quality«

The result of assessing all the recognized and nameable features and feature values of a speech sample under examination, in terms of its suitability to fulfil the expectations of all the recognized and nameable features and feature values of individual expectations and/or social demands and/or demands.   (cf. page 6)

»speech test«

A routine procedure for examining one or more empirically restrictive quality features of perceived speech with the aim of making a quantitative statement on these features.   (cf. page 91)

»successful quality design«

Elements make their mark in the quality shaping process of an entity in that their finished forms are offered to persons who perceive and assess quality. They assign quality features to the features of the entity, or its perceived sub-entities, that they deem to be identical to those they regard, in the sense of individual expectations and/or social demands, as desirable or positively expected in the entity and its sub-entities.   (cf. page 17)

»entity«

"Material or immaterial object under observation". [49]   (cf. page 16)

Subject Index

Aachen Logatom Test 100
absolute threshold 75, 79
acceptance 83, 101
accommodation 57
anticipation 54, 115, 134
approach
    diachronic 145
    synchronic 145
articulation 85, 101
    tract 29
aspects of invariance 56
assessment 64, 107
assessment scenario 7
    artificial V
    model-based 7
    natural .... V, 7
assimilation 57
attention 93
audiology 24, 25, 97–102
auditory system 99
automatic
    announcement 36
    concept-to-speech 36
    text-to-speech 36
Bellcore Test 98
black box 35, 106
calculation 64
call clarity index 66
categorical estimation 93
Category Assessment Test 101

causal sign 41
channel 24, 25, 44
clarity 77
cluster
    consonant 118–166
    frequencies 119
    identification 124
    structures 120
cluster identification
    CLID test 98, 118–129
    methodology 119–124
cluster-similarity study 143–173
cocktail-party effect 24
cognition 53–55
communication quality 49, 66
communicative signs 43
commutation test 145
comprehensibility 100, 113, 129
    tests 114–141
comprehension .31–32, 49, 53, 77–78,
                86, 93, 100, 129
    problems 101
concept-to-speech synthesis 36
conceptualization 16
constancy 56
    of attitudes 20
content words 133
conversion 21, 74
    grapheme-to-phoneme 46
    text-to-speech 36
correct value 78

correlation analysis 84
dendrogram 155–166
design 16
desired features 33
detecting
    equivalents 75
    the absolute threshold 75
    the differential threshold 75
diagnosis 107
diagnostic rhyme test 98
dialog-to-speech synthesis 36
differential threshold 75
distance relation 135
distinctive features 109, 131, 158, 159
distortion
    temporal 65
    time-variant 66
DRT 98
ease of communication 159
elements of synthesis 46
E-model 66
epistemology 8, 53
evaluation 107
event
    acoustic 60
    auditory 55, 60, 164
    auditory speech 58
    dynamic 8
    perception 14
    sound 71–74
    speech perception 61, 85
    speech quality 6
expectation 54
experience of technology 34
experiment 92, 93
field test 106
formant synthesis 46
frequency of occurrence 137
    English phonemes 99
    German monosyllables 139
    syllable structures 121
function words 133

Gaussian distribution 150
German Sentence in Noise Test 100
glass box 35, 106
Göttingen Sp. Intelligibility Test ..100
graphemic sequence 122
Harvard Psychoac. Sentences 99
Haskins Sentences 99
hearing aid technology 24
Hearing in Noise Sentence Test 100
hierarchical cluster analysis ..153, 158
immersion 19
individual differences 93
influence 130
    of cluster frequencies 127
    of cognition 93
    of context 93
    of interference 93
    of scaling 131–133
    of speaker characteristics 93
    of stimulus context 133–134
    of syllable structure 127, 134–138
    of test material 93, 124–129
    of word class 125
information carrier 2, 45
INMD 66
instrumental
    intelligibility measure 65
    investigative process 64
    measurement 64, 68, 71
    measuring method 65
    measuring procedure 106
intelligibility 31
internal reference system 93
Kandinski test 102
Kolmogorov-Smirnov test 150
laboratory test 13
language-in-use 47–49
listener dimensions 93
listening effort 66, 83, 101
listening effort scale 83
listening experiments 148
logical model 4

loudness 63
loudness-preference scale 83
LPC synthesis 46
magnitude estimation 80, 93
manually driven synthesis 36
market research 16
measurand 61–63, 77, 140
    of speech 77
    physical 60
measurement 63
    auditory 63
    design of 91
    instrumental 62, 65
    non-intrusive 66
    procedure 140
    quality of 89
    scale 59
    standardization of 90
measuring
    apparatus 65, 69, 71
    instruments 65–68
    object 3
    organs 65–68
    scale 80–86
    value 63
measuring object speech 89–102
methods
    of auditory measurement 65
    of psycho-physics 60
    of speech quality assessment .5–9
metrology 8, 60, 81, 113, 139, 148
model patterns 144
model types
    monitoring 66
    network planning 66
    signal-based comparative 66
models
    logical 4
modified rhyme test 98
MRT 98

names
    geographical 140
    given 140
natural communication 14, 25
natural speech ...15, 28, 43, 43–46, 48,
                                              85, 90
naturalness 77
network of relationships 8
non-parametric test 150
nucleus 118
Numbers Test 100
objective measuring process 67
objectivity 58
observation 64, 77
Oldenburg SpeechTest 100
overall impression 85, 101
paired comparison 93
parametric synthesis 46
parametric test 83
perception 53–59
    controlled 14
    random 14
phenotechniques 18–19
phon 63
phoneme-to-speech synthesis 36
phonetic similarity 150
planning phase 16
points of subjective equality 75
Postcard Test 102
procedure
    feature-oriented 114
process
    calculating 64
    instrumental 64
    investigative 64–65
    measuring 64–65
    statistical estimating 64
product quality 1, 11, 50
projection model 7–9, 110
pronunciation 102
PSOLA synthesis 46

psycho-acoustics 1–4, 8
   classical 78
   methods 75–80
   speech-related 78, 84, 148
psychometry 140, 150
psycho-physics 60, 75
quality 17, 82
   checks 11
   elements 17, 78, 86, 113, 114
   features 6, 11, 17, 78, 82, 91
   planning 11
   requirements 11
   test implementation 11
quality prediction 65
rating 149–153, 164
reaction time measurement 93
reading 37–41
reflection 6, 8, 14, 15, 55, 58, 74
response mode 94, 130, 141
SAM Overall Quality Test 101
SAM Segmental Test 116
scale
   CR-10 category 84
   function of measurand 80, 82
   standard 80
scaling 149–153
Semant. Unpredict. Sent. Test . 99, 117
semantic differential 82
semiotics 2, 45, 86
Short Conversation Test 102
similarity profile 153–166
SNR 130
sone 63
speaker recognition 77
speaking rate 102
Speech in Noise Sentence Test 100
speech intelligibility
   standard tests 97–102
speech intelligibility index 65
speech material 91
Speech Perception in Noise Test ... 100
speech perception processes 6

speech quality 6
   formal aspects 89
speech quality assessment
   general objectives 7
speech quality measurements
   factors of influence 93–95
speech test 91
speech velocity 77
Standard Segmental Test 97
structural concept 8
structuralistic approach 7
structure
   categorical 7
   static 8
SUBMOD-model 66
SUS 99, 117–141
Swedish Intelligibil. Sentences 100
synthesis
   time-domain 46
system quality 31
system theory 71–74
telephone speech 85
tele-virtual system 19
test stimuli 89
   presentation 91
text-to-speech synthesis 36
training 93
translated text-to-speech synthesis .. 36
transmission channel 36
true value 139
Two Syllable Rhyme Test 100
validity of test results 139
vocabulary 91
vocal tract synthesis 29
voice
   pleasantness 102
Ward method 153
word generator 120

Author Index

Adams E 84
Allen J 99
Altom M 98
ANSI 65
Argente J A 45
Aschoff V 29
Ausubel D P 57
Backhaus K 154
Bamford J 100
Bänsch D 100
Bappert V 93
Barnwell T P 65, 66
Barry W 97
Basson S 93
Beerends J G 65
Belhoula K 120
Bench J 100
Bennet R W 93
Benoit C 99, 117
Berger J 65, 66
Blauert J ... 6, 18, 36, 53, 60, 71–72, 93
Block von S 145
Bodden M 24, 93
Boogaart T 93, 94
Borg E 75, 93
Borg G 75, 83, 84, 93
Borg I 59, 60, 66, 67
Bosshart L 100
Brainerd C J 57
Brand T 100
Breuer S 32
Bronkhorst A W 24

Bronwen L J 83
Burkhardt F 28
Campbell N 24
Campbell W N 45
Cherry C 42
Childers D G 32
Chilla R 100
Clements M A 65, 66
Cohen M M 93
Conte S 93
Cosatto E 27
Dandi / Hcrc / Elsnet Workshop 107
Dedina M J 93
Delogu C 93
DIN 1319 .... 61, 64, 78, 139, 140
DIN 45621 Part 1/2 26
DIN 45626 26
DIN 55350 Part 11 .... 11, 12, 16
DIN 55350 Part 12 .... 12, 14, 17
DIN 55350, Part 13 64
DIN EN ISO 9000 13
Döring W H 100
Dudley H 29, 31
Dutoit T 23
Edwards A D N 48
Ehrenfels von Chr 56
Eisler H 83, 93
Ellermeier W 53, 81
Elliot L L 100
EN 60268-16 65
Endres W 31
Eskenazi M 45

Espesser R 93
Fabian R 56
Fant G 44, 47
Fechner G T 75
Fellbaum K 36, 90
Fetsen J M 93
Feustel P C 93
Flanagan J L 29
Foley H J 54
Fourcin A J 93, 97
Furui S 23, 24
Gabriel P 100
Gaitenby J 99
Galanter E 83
Garber E E 93
Gelfand 75
Gersho A 65
Gescheider G A 83
Gibbon D 91, 93
Gierlich H W 102
Ginsburg H 57
Goldstein M 93, 98
Goodman D A 93
Graf H P 27
Grave H F 80
Greene B 98
Greenspan S L 93
Gregory R L 54
Grice M 99, 101, 117
Gross S 37, 38, 41
Guilford J P 84
Guirao M 93
Guski R 53
Hagerman B 100
Hahlbrock K H 26, 100
Halka U 65
Hamacher V 100
Hansen M 66
Harland G 97
Hauenstein M 65, 66
Hauptmann A 36
Hazan V 93, 94, 97, 99, 117
Hecker M 98

Heidenfelder M 81
Heinrichs R 102
Hellbrück j 53
Henle M 56
Hermann T 89
Hess W 90
Heute U 29, 65
Hicks D M 32
Higuchi N 24
Hirose K 93
Hirst D 101
Hörmann H 40
Hottenbacher A 102
House A 98
Houtgast T 65
Howard-Jones P .93–94, 97, 101, 116–
                                        117
Hunnicutt M S 99
Iida H 36
Ingold L 100
ITU -T Suppl 3, P-Series Rec 66
ITU-T COM 12-101 66
ITU-T COM 12-C1-E 16
ITU-T Rec G.107 66
ITU-T Rec P.56 66
ITU-T Rec P.561 ....66, 85
ITU-T Rec P.562 66
ITU-T Rec P.800 81
ITU-T Rec P.861 66
Jain A N 36
Jakobson R 44, 145, 146
Jekosch U 2, 15, 16, 53, 57, 93, 98,
                        99, 109, 114, 117, 118
Jo C W 120
Jones B J 93
Jongenburger W 85, 95
Kabas M 100
Kalikow D N 100
Kasuya H 93
Kasuya S 93
Katz D 56
Keller E 30
Kettler F 102

Kim K T 120
Klatt D 99
Klaus H 90
Kliem K..100
Koffka K 56
Kohler K J 56
Köhler W...56
Kohonen T.......................................55
Kollmeier B 26, 66, 100
Koopmans-Van Beinum F 42
Kornatzki von P 38
Köster S..27
Kowal A 100
Kozielski P 100
Kraft V 45, 90
Kratzenstein C G............................29
Kröber G 8
Krumhansl C L................................93
Kryter K 98
Kurematsu A 36
Langhoff T 48, 49
Lankhorst E....................................93
Lazarus H 65
Lee Y J 120
Lenke N...42
Leontjew A A..................................48
Levas S 93
Lewandowski T.........................40, 46
Lewis E 24, 28, 32
Lieberman P 93
Lienert G A 86, 92
Lindström B 93
Ljunggren G 83
Logan J..98
Logan J S...93
Lucas J P 93
Luce P A..93
Luce R D 93
Luchins A S.....................................56
Luchins E H 56
Lucid D P 4
Lüschow 90
Lüschow F................................89, 90

Lüschow W 47, 48
Lutz H D 42
Macchi M 98
Malter B 123
Manous L M....................................93
Marks L E 83, 93
Matlin M W.....................................54
Mc Nair A E....................................36
McManus P R 83
Merleau-Ponty M 54, 55
Mersdorf J 27, 46
Messik S 84
Middelweerd M J 93
Mimpen A M...................................93
Minematsu N...................................93
Moles A A 19
Möller S 24, 27, 66, 102
Moore R 91
Morimoto T.....................................36
Morton K...32
Ney H 24
Nikléczy P.......................................30
Nilsson M......................................100
Nöth W 146
Nusbaum H C..............................93, 95
Nye P...99
Ohne S...93
Olaszy G..30
Opper S 57
Ostermann J 27
Ozawa K...93
Paoloni P 93
Pavlovic C 93
Pavlovic C V 65
Peckels J..98
Pfister B 24, 45
Piaget J 56, 57
Pijper J R de 36
Pisoni D B 53, 93, 95, 98
Plomp R 93
Pocci P 93
Poertele T 101
Pols L C W 109

Pörschmann Chr 19
Portele T 45, 90
Price D D 81
Quackenbush T P 65, 66
Raake A 24, 66
Raatz U 86, 92
Reitzle W 18
Remez R 53
Richards D L 102
Riesz R R 29
Rietveld A C M 93
Rinscheid A 32
Rossi M 93, 98
Rosson M B 93
Roumiguiere J P 93
Sagisaga Y 24
Sagisaka Y 24
Sahrhage J 18
Saito H 36
Saussure F de 3
Schaffert E 36
Schröder M R 29
Schröter J 27
Schubert K 100
Schulte-Fortkamp B 82
Schwab E C 93, 95
Seiler T B 146
Sekey A 65
Sementina C 93
Shi B 93, 94
Shikano K 36
Silverman K 93, 94
Simon C 93
Sixtl F 84
Smith B 56
Soli S D 100
Sonntag G P 101
Sorin C 93
Sotscheck J 90, 93, 98, 114
Speyer R 93
Spiegel M 98
Sprenger M 42
Stankowski A 37

Staufenbiel T 59, 60, 66, 67
Steeneken H J M 65
Stemerdink J A 65
Steven K N 100
Stevens S S 83
Stöber K 32
Styger T 30
Sullivan E V 57
Sullivan J A 100
Tarnoczy T H 29, 31
Tatham M 24, 28, 32
Tebelski J 36
Till O 93
Traber C 45
Trautmann U 48, 49
Tschopp K 100
Vagges K 101
Van Bezooijen R .. 85, 93, 95, 105, 106
Van Erp A 99, 117
Van Heuven V J 105, 106
Voiers W 98
Vollmer G 56, 58, 85, 148
Wagener K 100
Wagner P S 32
Wahlster W 36
Waibel A 36
Wallace K 98
Wang S 65
Ward L M 83
Warren H C 60
Watkins S S A 29
Waugh L 44
Weizsäcker von E 146
Wertheimer M 56
Wesselkamp M 65, 100
Westphal W 81
Williams C 98
Winski R 91
Wlater J 27
Wu K 32
Yegnanarayana B 32
Zimbardo P G 56
Zwislocki J J 60, 93